Coral Stone Mosques
of Maldives

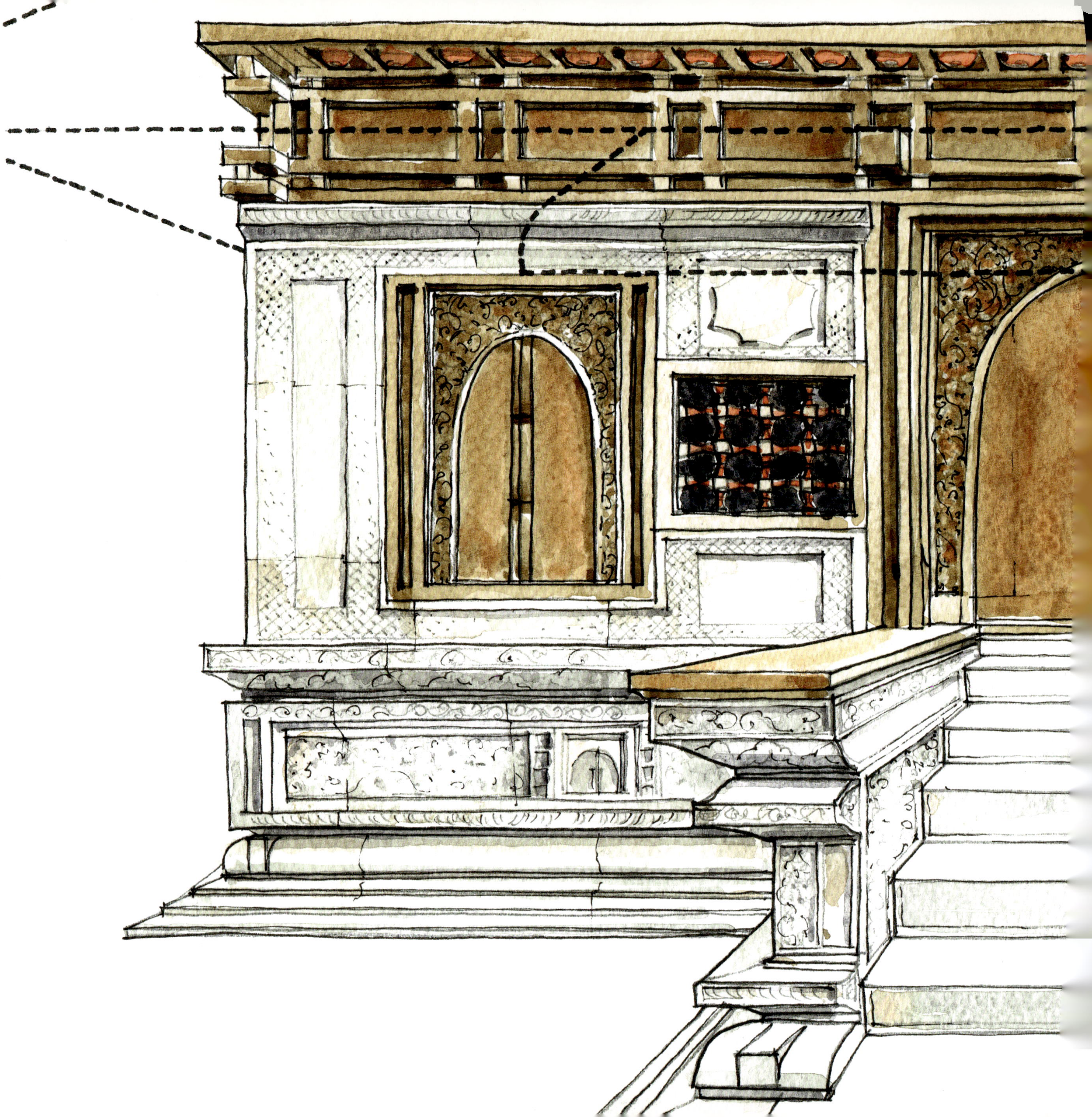

Coral Stone Mosques *of* Maldives

The Vanishing Legacy *of* the Indian Ocean

Mauroof Jameel and Yahaya Ahmad   Edited by James Steele   Illustrated by Mauroof Jameel

Gulf Pacific Press

First published in the United States of America in 2016 by
Gulf Pacific Press
983 West 30th Street
Los Angeles
California
90007
United States
00 1 212 677 3823
www.gulfpacificpress.com

ISBN: 978-0-986-28184-6
Library of Congress Control Number: 2015914462

Front Cover	Male' Eid Mosque without its roof structure
Back Cover	Fenfushi Old Friday Mosque, coral carving on the plinth
Pages 2–3	Fenfushi Old Friday Mosque without its roof structure
Page 4	Isdhoo Old Mosque without its roof structure
Pages 6–7	Male' Kalhuvakaru Mosque without its roof structure
Page 9	Male' Eid Mosque, the timber entrance door to the prayer hall
Page 10	Male' Old Friday Mosque, the central prayer hall
Page 11	Male' Eid Mosque, stair details
Page 13	Isdhoo Old Mosque, coffered ceiling
Page 14	Fenfushi Old Friday Mosque, the central prayer hall
Page 15	Fenfushi Old Friday Mosque, *Laage'*
Page 18	Ihavandhoo Old Friday Mosque without its roof structure
Page 19	Male' Old Friday Mosque, the central prayer hall

Designed and produced by ARTMEDIA, London

Printed and bound in Italy

Acknowledgements

We would like to acknowledge and express our heartfelt gratitude to the following people and institutions; without them this book would not have been possible:

- Mr. Mohamed Umar Maniku, Bandos Island Resort and other patrons of architectural heritage research of Maldives who chose to remain anonymous, for their generous financial contributions.
- The Heritage Department of Maldives, its former and present state ministers, directors and staff who facilitated the research.
- Members of the World Heritage Nomination Project survey team of 2011 (Ahmed Zameer, Ali Wiam, Ahmed Sharim, Mariyam Isha Azeez, Aishath Shuza Mohamed Rushdhee and Ibrahim Muja) for their systematic initial data collection and support.
- Professor Naoko Fukami for her invaluable advice on the history and cultures of the Indian Ocean, the Islamic world and research on coral stone mosques of Maldives.
- Madam Naseema Mohamed for her invaluable advice on the history and culture of Maldives.
- Ahmed Mauroof Jameel for his advice and contribution on the editing and design of the book.
- Ahmed Asad Mohamed and Hushan Shafeeg for their contributions to the design of the book during its early stages.
- Umair Badeeu, Arshad Jameel, Ahmed Ramie Rasheed and Imran Zaki for their comments on the design of the book.
- Research assistants from the University of Malaya and the International Islamic University of Malaysia (Faezeh Nabawi, Nadrahtul Aiman, Nurul Fariza and Asma-Dinie).
- Aminath Hassan for the initial telephone surveys in Maldives.
- Mohamed Ali (Moger) for his management and logistical support during all the stages of the book.
- Mohamed Ibrahim and his family from Ihavandhoo; Gasim and his family from Meedhoo; Dr Mausoom from Sun Island Resort; Shahid from Fenfushi; Mohamed Khaleel and his friends from Laamu Atoll; Ibrahim Easa from Isdhoo; Adnan, his family and friends from Vaadhoo; Adham, Anwar, Shiyama and her family and friends from Addu and all the people who provided hospitality and logistics during our travels in the islands.
- Our family and friends who generously supported us in every way, during all stages of research and compilation of the book.

We thank you all from the bottom of our hearts.
Mauroof Jameel and Yahaya Ahmad

Photo credits:
Mauroof Jameel
Ahmed Shinan
Ali Wiam
Mariyam Isha Azeez

In each island of theirs [Maldives] there are beautiful mosques.

Ibn Battuta, 1355

Contents

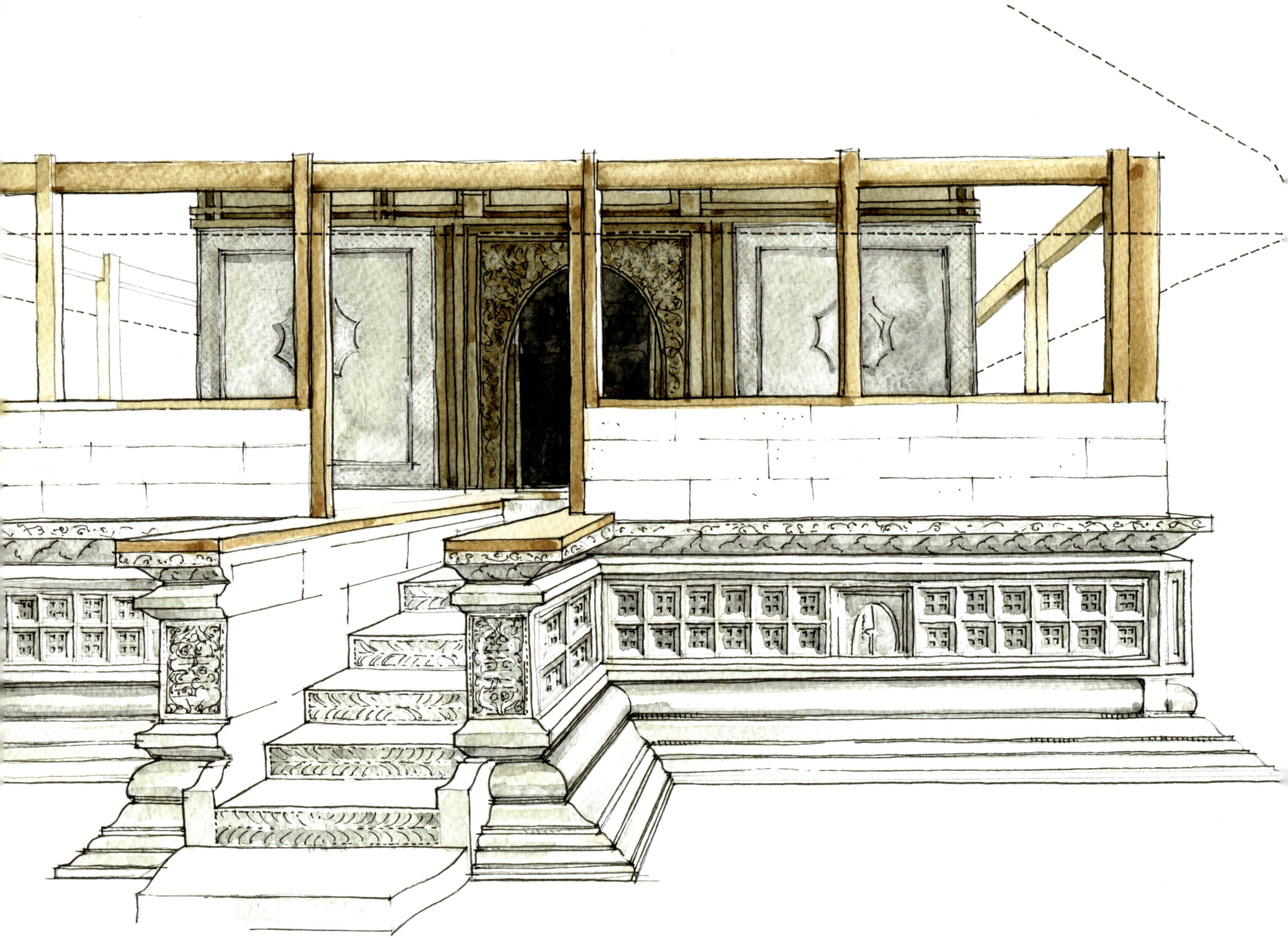

... these outstanding monuments [coral stone mosques of Maldives] have still not been appreciated by historians in recording the story of the world of Islamic architecture ... These are fine, unique examples of Islamic architecture and should be safeguarded at any cost.

Roland Silva, Former Commissioner of Archaeology, Sri Lanka, 1985

Preface

The coral stone mosques of Maldives are a type of tropical mosque built from ornamental interlocking coral stone (*hirigaa*) from the reef, using a construction technique called 'coral stone carpentry'. The mosques have intricately carved coral stone and detailed timber lacquerwork not seen in any other part of the world. Today the coral stone carpentry technique is extinct and coral mining is no longer possible for environmental reasons. The mosques that survive are the last of their kind and have become monuments of universal outstanding value that need protection for future generations. Roland Silva's quote opposite remains valid today; we still do not know enough about these mosques and many are in a poor state of conservation. Their beauty and elegance cannot be fully appreciated because most of them have been modified and substantially extended. The legacy of the coral stone mosques of Maldives is seriously endangered and needs appreciation for future protection and conservation.

previous spread
Male' Old Friday Mosque without its roof structure
opposite
Male' Kalhuvakaru Mosque, coffered ceiling and beam

This book is an introduction to coral stone mosques with illustrations of selected mosques in their former glory. It is intended to raise awareness among a variety of users, as a reference work that is simplified without compromising accuracy. This book originated from recent research work at the University of Malaya and a UNESCO World Heritage Nomination Project for the government of Maldives.

The book is presented in three chapters. The introduction chapter provides a background to the cultural heritage of Maldives, introduces coral stone construction, establishes styles and features of mosques and identifies the cultural and architectural influences from the region on coral stone mosques. The second chapter describes the traditional mosques of Maldives and lists those noted by the authors. The final chapter identifies the typical features, and eight selected mosques, to tell the story of coral stone mosques in their present condition and past splendour. The book concludes with an appendix covering a brief historic background and a historic timeline.

Introduction

Cultural Heritage

We are justified in ranking the culture of the Maldives as a civilisation, which is technically defined as having urbanisation, writing, calendars, structured law, a state system, substantial buildings and specialists in such areas as religion and trade.

Clarence Maloney, South Asian anthropology scholar, 1980

The culture and life of the traditional Maldive islands is testimony to the human genius for adaptation and survival among small islands in the middle of the ocean. Maldives has been inhabited for more than 4,000 years and the primer of its culture is its long experience in survival with limited resources in the middle of the ocean and interactions across the ocean. A nation where 99 per cent of its area is sea, the livelihood of its people has totally depended on the sea, developing a unique maritime culture initially trading with cowry shells, coir rope, tortoise shells and ambergris and today, in modern times, trading with fresh fish and tourism.

Maldivians learned the patterns of the ocean, making a calendar system around the weather, building specialist boats, and mastering sailing and fishing techniques, fish preparation and cuisine to suit their maritime culture. Even their performing arts and stories were related to the sea. From the beginning, Maldives was also influenced by travellers and coastal civilisations of the Indian Ocean. Maldivians played a significant role in Indian Ocean travel and trade during ancient times, as accomplished travellers trading their sea products and as providers for traders and travellers who passed through the area.

With the increase of travel and trade in the Indian Ocean region, people from different cultures in the region began to settle in Maldives. Thus Maldives unique maritime culture evolved with influences from India, Sri Lanka, Arabia, Persia, Eastern Africa and the Malay Archipelago all the way up to China. The cultural heritage of Maldives consists of an intangible heritage based on the unique language, performing arts, arts and crafts, and cuisine and a tangible heritage based on movable properties such as museum and antique collections and immovable properties based on pre-Islamic and Islamic archaeological sites, underwater archaeological sites and architectural heritage.

Language and traditional performing arts

previous spread

View of a typical traditional island settlement

The unique language Dhivehi, spoken only in Maldives and on Minicoy Island, provides Maldivians with their strongest cultural identity. It is based on Elu, an ancient form of Sinhalese influenced by Sanskrit, Pali, Persian, Arabic and Urdu. The language is written from right to left using the Thaana script that originated around the 1600s, succeeding two other scripts called Eveyla and Dhives.

The traditional performing arts of Maldives, which include songs, music, storytelling and festivals, are derived from the region and its seafaring cultures. Cultural songs include *raivaru*, sung poetry with swapped letters; *bandhi*, rhyming poetry or chants; *farihi*, ballads; and *ambaa*, repetitive chants or shanties related to work such as rowing. Cultural songs also include religious songs, such as *madhaha* and *mauloodh*, expressing love for and narrating the life of Prophet Mohamed (peace and blessings be upon him). The cultural music is mainly based on *boduberu*, *thaara*, *langiri* and *bandhi*.

Boduberu drum music is usually played with three double-headed drums, percussion, a lead singer and a chorus. The music generally starts with slow drumbeats, chanting songs, percussion and hand clapping, working to a climactic fast beat while accompanying dancers slowly build up to a frenzy of animated movements. *Boduberu* music originated from East Africa – as evidenced by the Swahili words in the songs.

Thaara music is usually played with an ensemble of single-headed frame drums very similar to *tar* drums found in North Africa and the Middle East. *Thaara* is performed with 22 people in two rows and starts as a slow beat working to a climax of fast beats, movements and chanting songs. *Thaara* music originated from the Middle East and North Africa as indicated by the use of many Arabic words.

Langiri and *bandiyaa jehun* are both music with rhythmic dance movements. They are usually performed in groups with melodic songs, drums and percussion in the form of sticks and pots. The music and dances are performed by both men and women to express joy during festive occasions. Most of the music and dance originates from Indian folk music and dances.

There are other performing arts such as storytelling and performances related to festivals and communication. Storytelling is no longer practised, although a lot of the ancient history has come down through stories. The story of Boduthakurufaanu was related by one of the most famous storytellers, Buraaru. There are performing arts related to different festivals such as royal festivals and religious festivals.

A double-headed drum used in *boduberu* music; a single-sided frame drum used in *thaara* music and small brass cymbals also used in traditional music

Kulhi jehun, a display of traditional martial arts, and *beru dhunmaari thaalhafili*, trumpets and drums played during royal processions, are both arts related to royal festivals. *Bodu maali neshun* and Eid games are performances and games related to religious festivals. Drums were used to communicate messages to the public. Special drums for public gatherings called *havaru beru*, and drums to mark midnight called *dhamberu*, are two examples of such well-known drum practices.

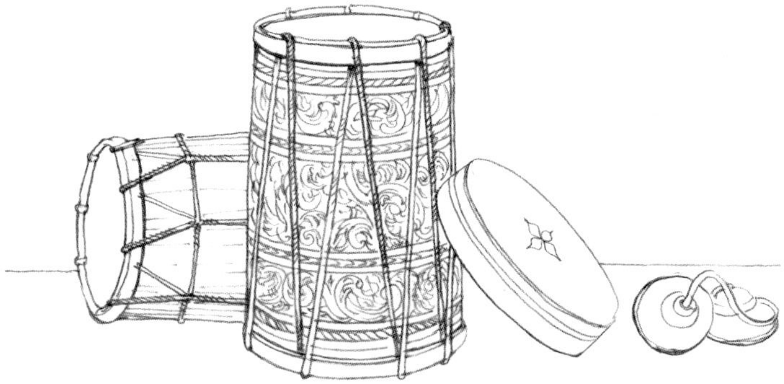

Traditional arts and crafts

The Maldivian traditional arts and crafts include coir rope work, lacquerwork, stone carving, mat weaving and plaiting, textiles and embroidery, metal work, recreational and ceremonial crafts and boatbuilding. From a very early period, visitors have admired Maldivian craftsmanship and skills. Coir rope, a product handmade from coconut fibre, was a famous Maldivian product during ancient times. It was valued and sought after because of its strength and durability and was mentioned by early travellers and scholars of the region as far back as 362 CE.

Liyelaajehun (lacquerwork), traditional mat weaving, *kasabu* (traditional embroidery) work and weaving are famous Maldivian crafts. Such products included in ancient trade, annual tributes and presents to foreign lands can be dated back to 362 CE. The weaving and plaiting of coconut leaves are also important crafts. Products such as *mulhi* and *gonu*, woven coconut palm leaf baskets used for carrying chickens, fruits, vegetables and other items, were traditionally used in everyday life.

opposite

Traditional lacquerwork

right

Design of a traditional mat: *thundu kuna*

far right

Stages in making coir rope from coconut fibre

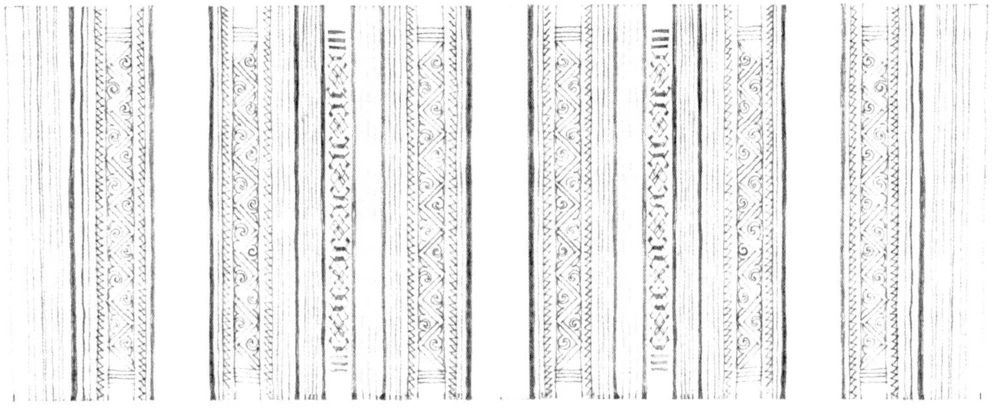

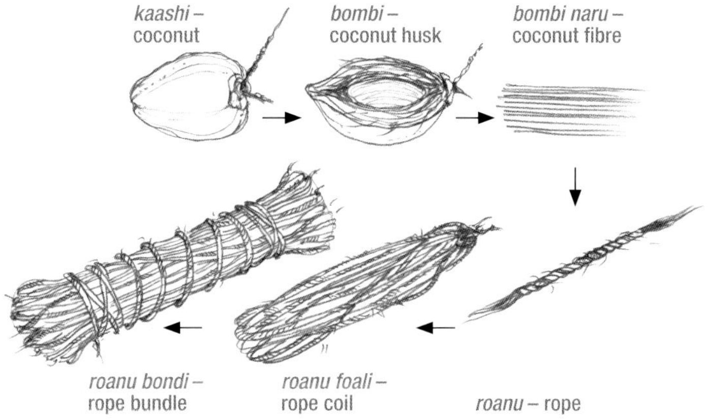

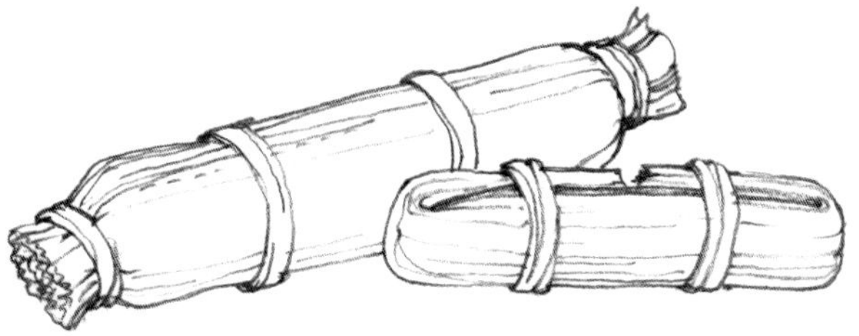

Traditional cuisine

Traditional Maldivian cuisine was mainly fish and coconut-based along with yams and breadfruit. Maldivians mastered the preparation of preserved foods to suit their long-distance travel needs and seafaring culture. This has resulted in many varieties of special preserved foods such as Maldives fish, dried tuna, *rihaakuru* (a thick fish paste), *dhiyaa hakuru*, *karu hakuru* and *fathuli hakuru* (types of coconut treacle), and *bondi* (a sweet made from coconut and coconut treacle), all of which are popular and traded in neighbouring countries such as Sri Lanka. With the increase of cultural exchange, Maldivian cuisine was influenced by the culinary arts of South India, Sri Lanka and the Malay Archipelago.

opposite

Bondi and *fathuli hakuru*: sweets made from coconut and coconut treacle

right

Dhivehi keyn: special ceremonial meals including rice and curries

far right

Hikimas or Maldives fish: smoked and dried tuna

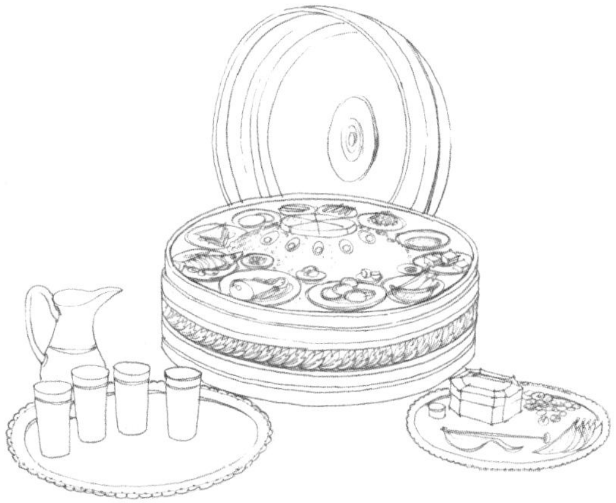

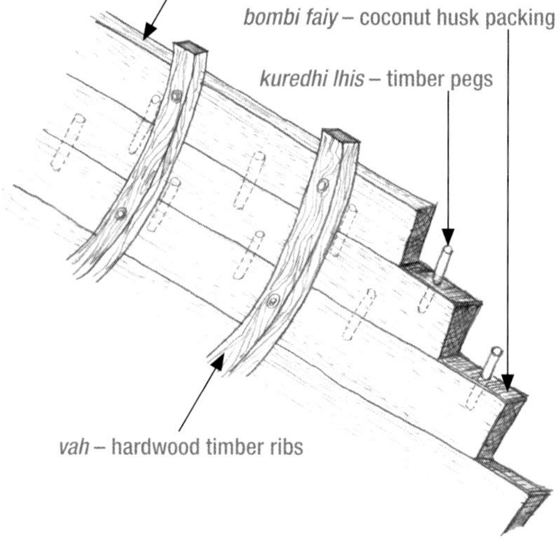

Traditional boatbuilding

Maldivians mastered the art of boatbuilding from very early times. Their work was admired by many travellers and scholars who visited the region: Al Idrisi in 1150, the famous traveller from Morocco Ibn Battuta in 1343 and the great Chinese mariner Zheng He's chronicler, Ma Huan, in 1425 are among those who admired the early sewn boats of Maldives. The word *dhoni* resembles the Malayalam word *thoni* but, unlike the *thoni*, Maldivian boats are planked boats similar to the dhows of the African and Arab regions bearing closest resemblance to the dhows of East Africa.

The first boats were planked boats sewn with coir rope and using woven, square sails. Later, exposure to changing boatbuilding techniques in the region initiated progress to metal rivets and lateen sails. Maldivian boats are well built and differ in size according to their use. The *bokkura* is a small light boat used for rowing and moving around the shallow reef flat and lagoon; a *vadhu dhoni* is a small sailboat used for fishing and transport closer to the islands; *mas dhoni* and *mas odi* are fishing boats with sails used for open sea fishing; *dhathuru dhoni* and *odi* are larger boats for inter-island travel and carrying cargo and *dhathuru battheli* and *bandu odi* are even larger boats for long-distance travel and cargo.

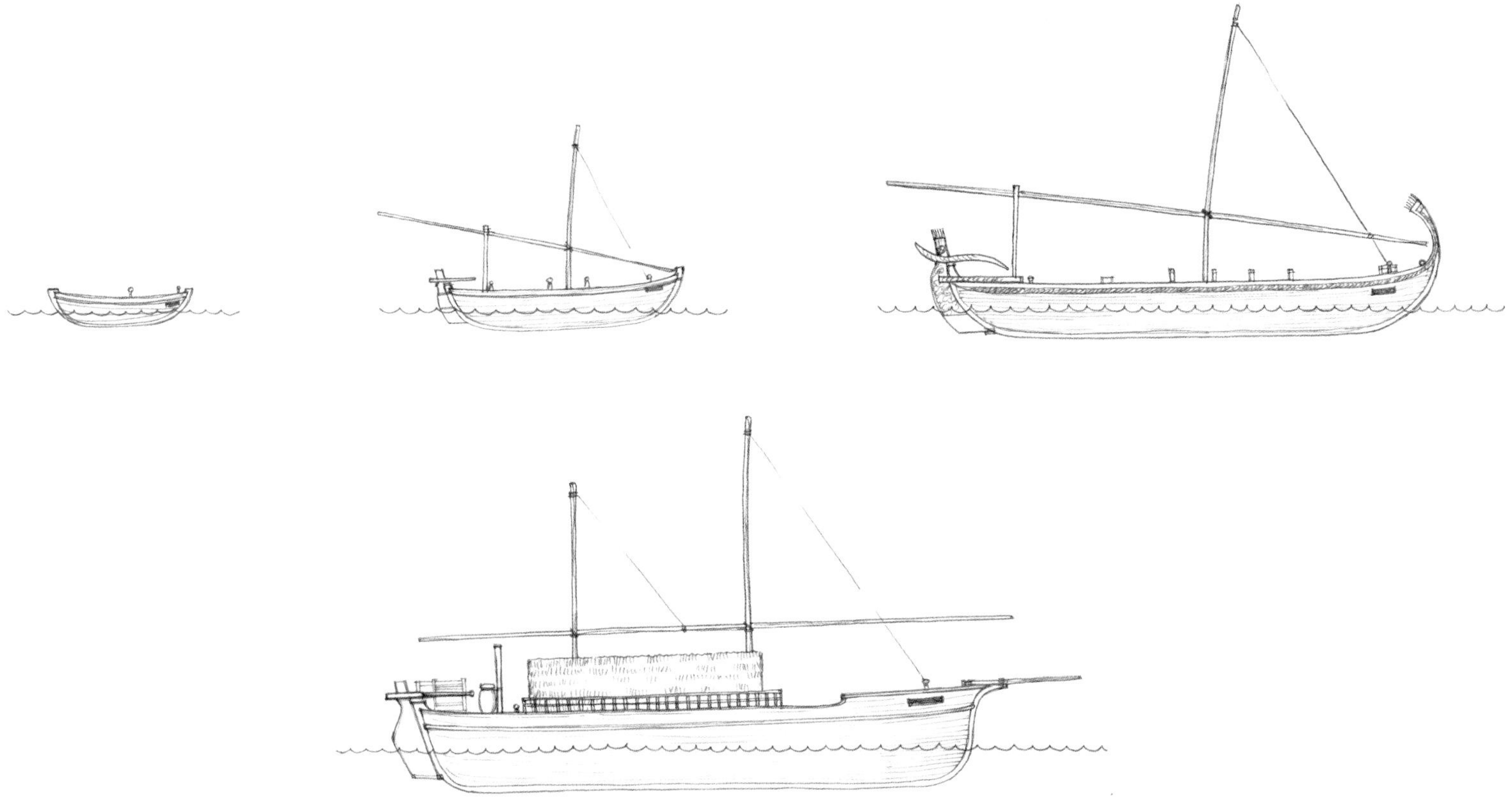

opposite
Detail of a traditional boat showing plank-and-peg boatbuilding techniques

clockwise
Bokkura: a small light boat of around 4.5 metres; *Vadhu dhoni*: a small sailboat of around 9 metres; *Mas dhoni*: a fishing sailboat of around 15 metres; *Dhathuru battheli*: a large sailboat of around 22 metres

Pre-Islamic cultural heritage

The immovable tangible cultural heritage of Maldives consists of pre-Islamic and Islamic archaeological sites, underwater heritage sites and architectural heritage.

The first Maldivians did not leave any archaeological evidence, as their structures would have been made from perishable materials such as wood and coconut thatching. The first people to leave physical evidence of an organised culture were the early Buddhists and Hindus when they started using coral stone structures. Despite limited literature on the subject, there are 78 known pre-Islamic archaeological sites and other evidence leading to a probable 1,700-year-long Hindu and Buddhist period, which has a foundational importance to the cultural heritage of the country.

opposite

Animal motifs found in pre-Islamic structures in Maldives

right

Thoddoo Dagaba, Thoddoo, Alif Alif Atoll, which has similarities to dagobas in Sri Lanka

far right

Haddhunmathi Kuruhinna Dagaba, Gan, Laamu Atoll, which has similarities to ancient dagobas in India

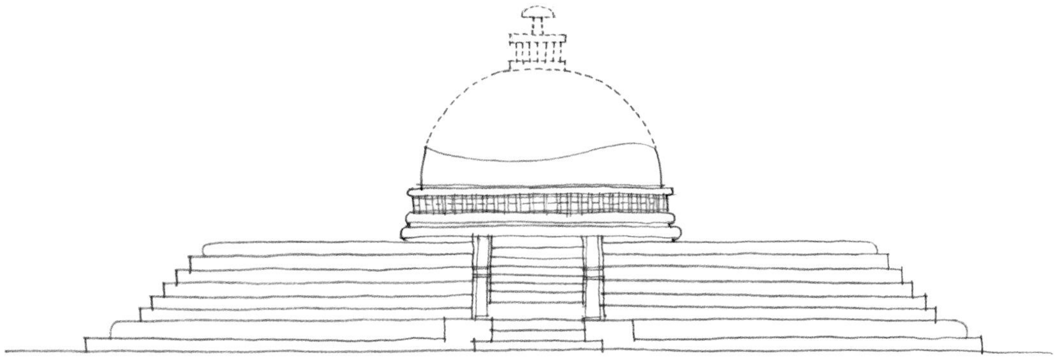

Vernacular architectural heritage

The architectural heritage that survives today includes vernacular architecture, traditional historic buildings such as mosques and few other structures or buildings. The vernacular architecture includes traditional island settlements and traditional houses, specifically for their forms and building techniques.

As the vernacular structures are often built using perishable materials such as timber and coconut thatch, they do not survive for long periods, but the island morphology, building forms and construction techniques have evolved through many centuries of experience. This experience encompasses responding to climatic conditions, available materials, changing technology and changing socio-cultural needs, including the religions of different periods. So in vernacular architecture, the ancient building forms, building principles and construction techniques are as important as surviving structures.

The traditional islands are built natural landscapes that illustrate a time-honoured way of life and have evolved through harmonious interaction between people and nature. The characteristic layout of a typical traditional island has an entrance and landing area for boats and a main road lined with houses and other structures. The entrance or landing area on the beach has *dhoni haruge'* (boat houses) and *holhu ashi* (small beach pavilions). The typical buildings of the island include traditional houses, mosques and *haruge'* (structures for purposes other than living).

Common features of a traditional island settlement:

a sheltered lagoon area

b boat landing beach

c boat houses (*dhoni haruge'*)

d beach pavilion (*holhu ashi*)

e main road

f mosque and its compound

g houses

h woods

A typical traditional
timber/thatch house

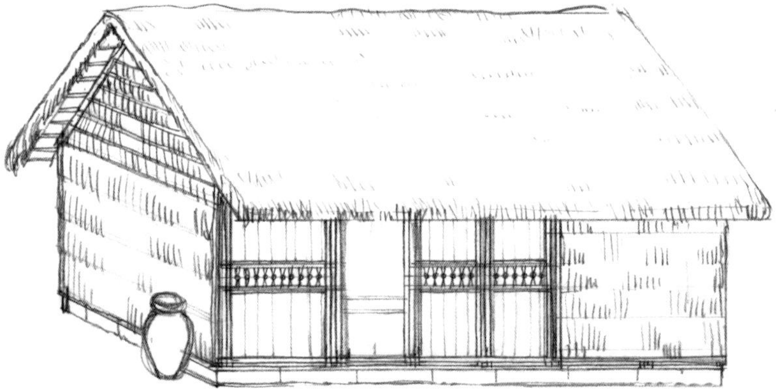

The outer chamber of a typical traditional house:

a multipurpose wooden platform (*bodu ashi*)

b coir rope bed

c wooden box

d entrance to inner chamber

e doorway to main backyard

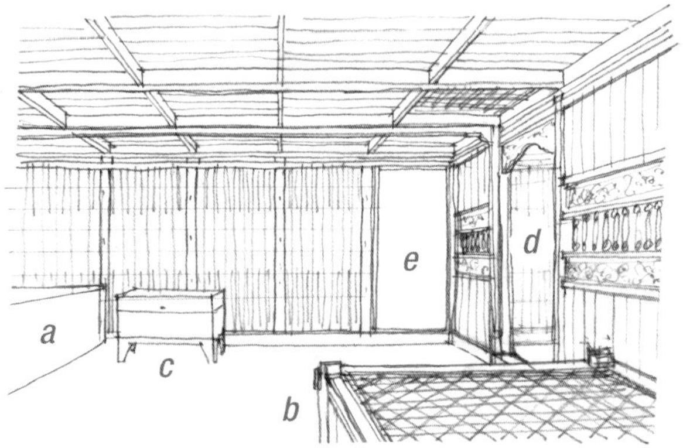

Traditional houses are located within compounds consisting of the house, kitchen, granary, well area and a toilet area. The older house compounds did not have a boundary wall separating them from other houses. A traditional house typically has two spaces, the outer chamber and the inner chamber. The outer chamber is the more public space entered from the front, while the inner chamber – entered through the outer chamber and a back door – is a private

Sectional detail of a typical traditional timber/thatch house showing traditional elements:

a coconut thatch
b battens (*fathi*)
c rafters (*faidhoh*)
d tie beam (*faiyfoshi vakaru*)
e beam (*bandu vakaru*)
f standing posts (*mella*)
g timber board (*filaa*)
h coral stone (*hirigaaolhi*) floor beam

Vernacular construction details of a typical traditional timber/thatch house:

a coconut thatch
b screwpine leaf mats
c stick lattice (*dhumburi*)
d coconut timber posts
e timber board
f coral stone floor beam

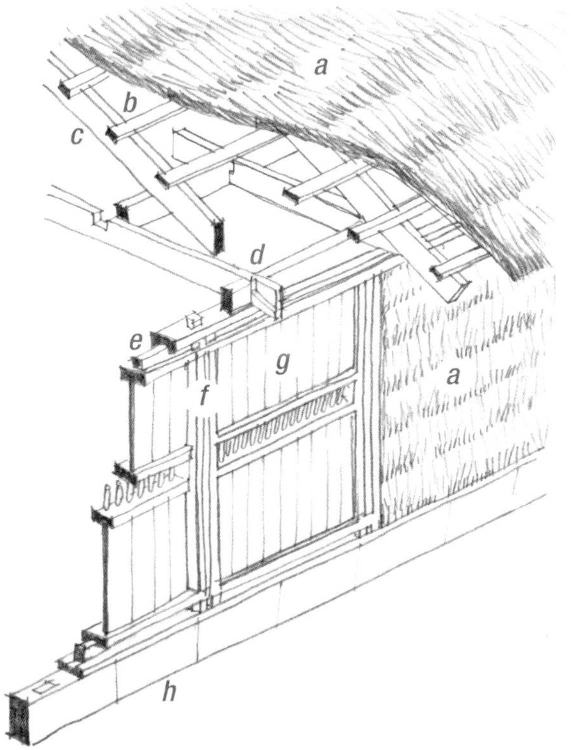

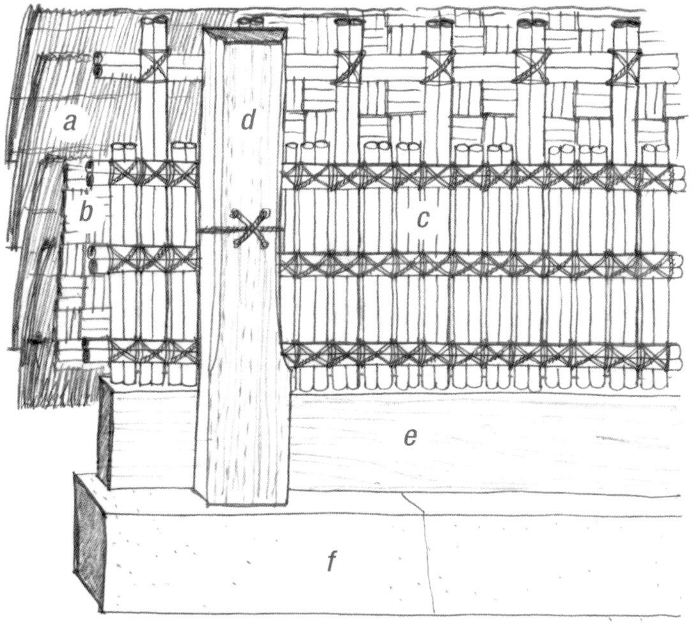

space occupied by the women of the house. Sometimes houses have open verandas at both the front and the back. Early housing was made of coconut thatch, or, depending on the prosperity of the owner, this might be combined with timber. Later housing was built from coral and lime masonry with a roof of clay roof tiles or corrugated iron sheets. The original form did not alter with the changes in materials and technology.

Coral Stone Construction

Corals reefs are found within a narrow band of low latitudes, roughly confined to the tropics and concentrated in the Indo-Pacific regions from the Red Sea, East Africa, and Southern Asia to the Central Pacific. There is a wide variety of corals but the most important that builds the reef – the stony coral family Scleractinia – is commonly used for construction throughout the coastal settlements of the tropical belt wherever it is found in abundance. It was used in Central American regions, the Arabo-Persian Gulf, the Red Sea, the Pacific and coastal settlements of the Indian Ocean. Today the best examples of coral stone architecture are seen on the Swahili coast of East Africa and Maldives.

The origin of coral stone construction is not well understood, though early use of coral in construction has been found among the Mayan island communities of Central America. Excavations in the Port Honduras region of southern Belize found construction using coral structures dating between 1500 and 900 BCE. Early use of coral in construction has also been found on the coasts of the Red Sea dating back to the Hellenistic period between 323 and 146 BCE. At the site of al-Rih in Sudan, a Hellenistic cornice made from coral stone was found reused in an Islamic tomb.

From the Red Sea, the use of coral stone spread to the Arab and East African coasts where it thrived as the primary construction material of monumental buildings for a long time. The best examples of coral stone construction in the Swahili coastal cities of East Africa are the buildings of the ancient trading towns of Lamu and Gedi in Kenya; Fort Jesus in Mombasa, Kenya; the Stone Town of Zanzibar in Tanzania, and the ruins of Kilwa Kisiwani and Songo Mnara in Tanzania. Lamu Old Town, a World Heritage Site, is one of the best-preserved Swahili settlements in East Africa using coral stone construction. The first stone mosques excavated in Shanga in the Lamu region, built during an estimated period of 900 to 915 CE, were constructed of neatly shaped Porite coral, bonded with mud with a white plaster face.

Coral stone construction also existed in Indian coastal cities, Sri Lanka, Maldives and some Indian Ocean islands. It was found at ancient Buddhist sites in Jaffna in Sri Lanka, in Maldives and in the Indian subcontinent as far back as the second century. Today the best examples of many types of coral stone structures are found in Maldives.

Comparing the construction techniques of East Africa and Maldives, both East Africans and Maldivians have structures that use coral stone set in lime mortar, plastered and lime washed, but the Maldivians used assembled, shaped coral blocks with an interlocking system and intricate carvings, a construction technique called coral stone carpentry. This type of coral stone construction is not found in any other part of the world. Coral carvings similar to those in Maldives can be found in Swahili coastal sites such as Kilwa Kisiwani and Songo Mnara, but they lack the variety and sophistication of detailing found in their Maldivian counterparts.

Coral stone construction in Maldives

Coral stone was the primary material used for monumental buildings until recent times in Maldives, as it was the most durable available. Coral stone construction in Maldives existed as far back as 165 CE during its classical Hindu and Buddhist period. Different construction techniques evolved responding to the means and needs of different historic periods. Today coral stone mining is not possible in Maldives for environmental reasons and many of the traditional techniques not found anywhere else in the world have disappeared, marking the end of a significant part of the cultural history of the country.

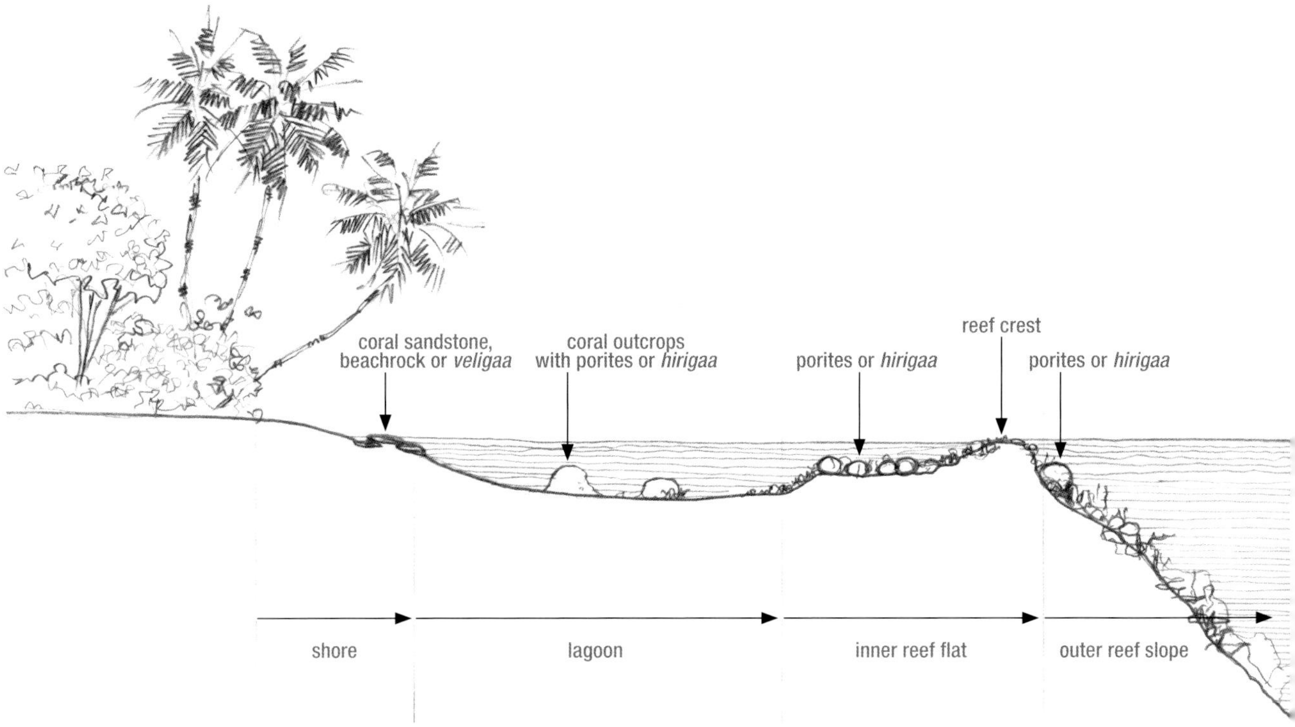

Typical section of an island reef showing the location of coral sandstone (*veligaa*) and Porite coral (*hirigaa*)

Early coral stone construction techniques probably arrived from Sri Lanka and India along with the Buddhist and Hindu cultures. Archaeological studies of the Kaashidhoo site, completed in 1998 by E Mikkelsen, revealed that shaped and decorated coral stone blocks set in mortar were used in the construction of the Buddhist monastery and the shell deposits found there were dated at between 165 and 345 CE. HCP Bell, on his archaeological expeditions in the Maldives, identified many Buddhist structures using shaped and decorated coral stone blocks set in mortar. Bell traced some of the features and profiles of the raised coral stone platforms of ancient Buddhist structures in Maldives to Sri Lankan Buddhist origins in the Toluvila Monastery in Anuradhapura.

Example of coral stone construction from the classical Buddhist period. Construction detail from the Buddhist site of Haddhunmathi Pirivena, Laamu Gan:

a compact earth floor
b shaped coral stone set in lime mortar
c lime skim coat

Two types of stones from the reef, coral sandstone and Porite coral, were used for construction in Maldives. Coral sandstone called *veligaa*, and sometimes called beach rock, is a coastal sedimentary formation consisting mainly of coral sand, lithified through the precipitation of carbonate cements. It is easily accessible, where available, but not suitable for decorative work due to its coarse formation. So it was useful only for mouldings and less intricate carvings. This type of stone was used during the early periods or in areas where the finer Porite coral stone was not easily available.

Porite corals, called *hirigaa*, from the stony coral family Scleractinia, were the main type of coral stone mined for construction in the Maldives. These boulder-like corals are slow growing, and coral heads bigger than one metre in diameter are commonly found among the reefs. They are also among the most important reef building corals and tend to dominate the back and lagoon areas of the reef flats. Traditionally, these Porite corals were mined for construction. Live coral boulders were extracted manually from the shallow parts of the reef, where it was available in abundance, using rods and ropes. It was then put onto a raft or boat and taken to land, chipped or shaped and dried before use. Techniques used in coral stone construction varied during different historic periods of Maldives.

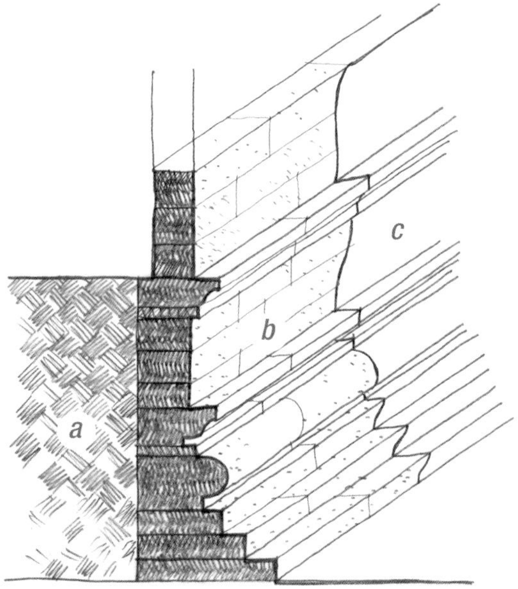

Example of coral stone construction from the medieval Islamic period. Construction detail of Gen *Miskiy*, Fuvah Mulah:

a compact earth floor

b shaped coral stone (*veliquaa*) set in mortar

c lime skim coat

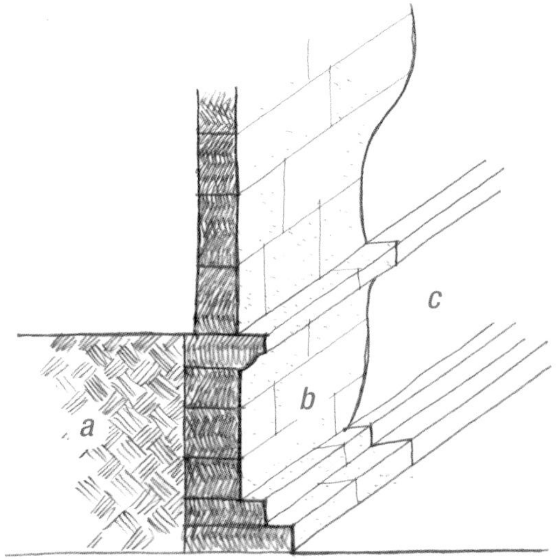

During the classical Hindu and Buddhist period (*c*500 BCE–1153) coral stone was shaped, set in mortar and finished with a lime skim coat. The structures had mouldings and a few decorative carvings, but the finishes were less refined. Both *hirigaa* and *veligaa* were used during this period. Archaeological surveys by HCP Bell in 1920, Thor Heyerdahl in 1986 and Egil Mikkelsen in 2000 found coral stone artefacts and coral structures of Buddhist and early Hindu origin. Examples of pre-Islamic coral stone construction are found in archaeological sites such as Kaashidhoo, excavated by Mikkelsen and the Laamu Gan sites excavated by HCP Bell.

During the medieval Islamic period (1153–1514), coral stone was shaped, set in mortar and finished with a lime skim coat. Both *hirigaa* and *veligaa* were used and the structures had mouldings, but no decorative carvings have been found from this period. Examples of Maldivian *miskiy* (mosque) structures from this period are the Boadha *Miskiy* in Koagannu, Hulhumeedhoo built around 1403, and Gen *Miskiy* in Fuvah Mulah, believed to be built around the 13th century.

Example of coral stone construction from the early colonial period. Construction detail of Male' Old Friday Mosque:

- a compact earth floor with screed
- b shaped coral stone assembled without mortar using carpentry techniques
- c coral stone surface decorated with carvings

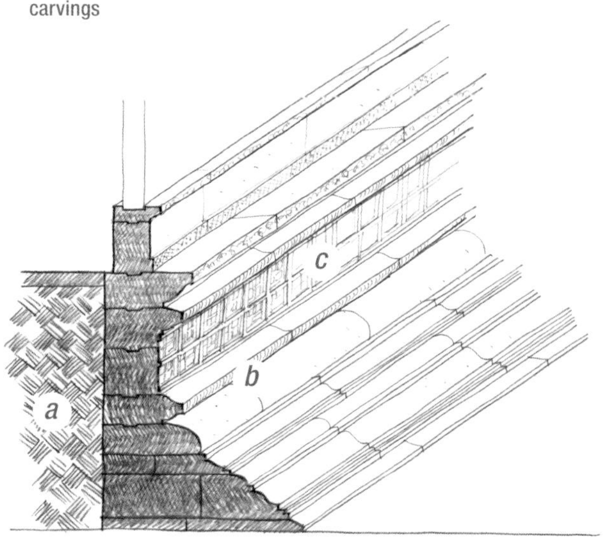

During the early colonial period (1514–1828), coral stone construction became refined as a result of increased travel and the exchange of ideas and skills in the region. Boatbuilding skills and construction techniques of the period fused with existing techniques from earlier periods leading to a sophisticated interlocking assembly technique without mortar.

Appropriately, the Dhivehi words '*hiri galu vadaan*' for this type of construction technique means 'coral stone carpentry'. Coral stone structures of this period had highly refined coral carvings and surface ornament. In 1617 Francois Pyrard of Laval gave an early detailed description of coral stone construction. The French traveller described a mosque in detail, and admired the skill of the Maldivian people in coral mining and preparation of the corals for construction.

Types of joints used in coral stone carpentry techniques:

a tongue-and-groove joint

b dovetail joint

c dowel-and-peg joint

d lap joint

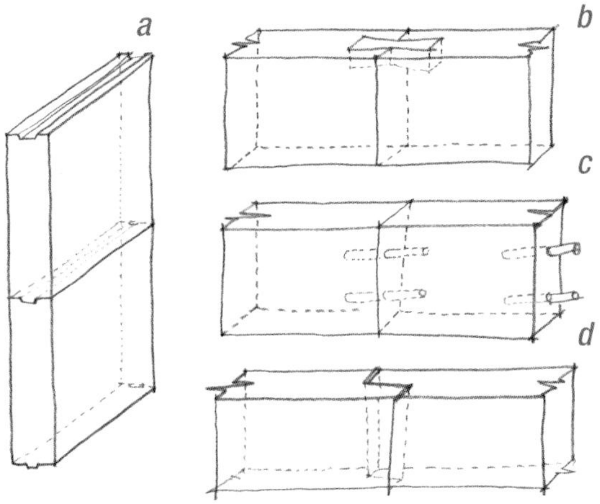

Hirigaa was the kind of coral used during this period, and three basic types serving different purposes have been identified. These are known as *onu hiri* (a coarse and grainy stone), *fuh hiri* or *oman hiri* (a smooth grain stone) and *goh hiri* (a knotty grained stone). For example, the type most suitable for carvings and finishing is *fuh hiri* or *oman hiri*. For the construction of mosques, large coral heads of suitable types were carefully chosen, crudely shaped into blocks and dried. The crudely shaped blocks were then transferred to the construction site, shaped precisely, assembled using carpentry techniques and finally decorated with carvings. Simple manual carpentry tools such as chisels and axes were the only tools used for the work. Examples of early colonial period coral stone construction are Male' Old Friday Mosque (1657) and Fenfushi Old Friday Mosque (1692–1701). Today there are 20 coral stone mosques in Maldives surviving in differing states of authenticity and integrity. They are the subjects of this book and are representative of the coral stone mosques of Maldives.

During the British colonial period (1828–1965), new coral stone construction techniques emerged as a result of changes in the dynamics of travel and the exchange of ideas and skills in the region. The earlier coral stone carpentry techniques slowly disappeared giving way to faster and more cost-effective techniques. Masonry techniques using chipped coral with lime mortar and plaster became the choice of construction for mosques and other monumental buildings during this period.

Many types of stony corals were mined from the sea and chipped into pieces and dried. Corals of the stony coral family were mined and burned to make lime. The coral pieces were set in lime mortar and finished with lime plaster and lime washed. The types of ornament use different mouldings on the plasterwork. Examples of construction techniques in this period include Bihurozukamana *Miskiy* in Male', built in the 1920s. This type of construction continued in Maldives until cement was introduced in the late 20th century, when coral pieces set in cement and sand mortar became the primary construction method.

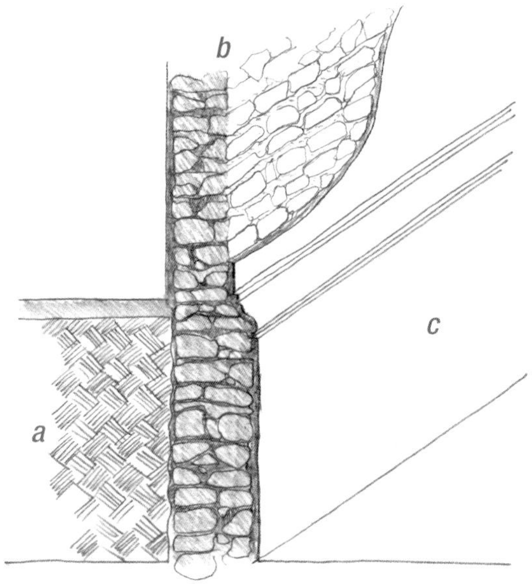

Example of coral stone construction from the British colonial period. Construction detail of Male' Bihurozukamana *Miskiy*:

- a compact earth with screed
- b chipped coral set in lime mortar
- c lime and sand plaster finished with lime white wash

As construction increased with the development of the country, the demand for coral stone also increased. Coral debris was being used as aggregate for reinforced concrete and in 1988 it was estimated that around 14,000 metric tons of coral were mined for use in the capital Male' alone. Coral and sand mining declined after mining from house reefs of inhabited islands was banned in 1992 and today coral stone construction is no longer possible.

The legacy of coral stone construction has become an indispensable part of cultural history, defining the creativity of Maldives. Coral stone construction using carpentry techniques survives only in the form of a few mosques and an unknown number of mausoleums, water wells and tombstones. Many of these mosques are in a poor state of conservation and often remodelled without due consideration for the historic and cultural value. The mausoleums and tombstones are neglected and not maintained. These coral stone structures need more appreciation and protection for the sake of future generations.

Mosque Architectural Heritage

The first mosque of Prophet Mohamed (PBUH)

The first house of prayer built when the Prophet Mohamed (peace and blessings be upon him) migrated to Medina in 622 CE was a simple shelter with palm tree trunks, supporting a roof of palm leaves. It was within his residential compound and was the place for meetings, congregational prayer and also the school, the courthouse, and the centre of Muslim life. After the death of the Prophet, the first four caliphs continued to use the mosque in the tradition of the Prophet, and the mosque structure remained simple and humble.

During the Umayyad dynasty that followed the first four caliphs, the role of the mosque changed from being the main centre of Muslim life to having a more specialised purpose. Different mosques were used for different purposes. Rulers separated their residences from the mosque, the madrasah-mosque emerged, the Sufi masters established separate mosques and bigger mosques originated for congregation and political status. The Abbasid dynasty that followed brought more developments to the typology of the mosque, many being the result of the concept of Islam as defined by the Abbasid caliphate and the Shia. They separated many traditional mosque functions. The palace or the governor's residence became the political and administrative centre, the madrasah became a separate educational centre, special courthouses were set up to administer the law, and so the mosque became the place of prayers, congregational sermons and public announcements. The Abbasids also started building tombs and mausoleums inside and near mosques, and began erecting palaces with a grand mosque adjacent, to symbolise the central authority in Muslim cities.

Today the main purposes of the mosque are as a place of congregational prayer and as a centre for religious learning, preaching, religious celebrations and religious administration. Mosques are found in three sizes that primarily respond to the needs and levels of prayers. First is the individual mosque or masjid. This is a small mosque used for daily prayers by individuals or small groups of people. The second is a Jami masjid or Friday mosque. This is a larger mosque used for community gatherings like the Friday congregational prayers and sermons. The third is the Idgah masjid or the grand mosque, used for large gatherings of whole towns and cities for prayers and sermons like the Eid prayers.

Typical regional architectural styles for mosques
The Central Arab lands

Mosques do not have a specific architectural style but as Islam spread to other regions of the world, different styles of mosque architecture evolved from the following regions:

1 The Central Arab lands
2 Spain and North Africa
3 Iran and Central Asia
4 Anatolia and the Ottoman region
5 The Indian subcontinent
6 West Africa and the Sub-Sahara region
7 East Africa and eastern Arab lands
8 The Chinese region
9 South East Asia

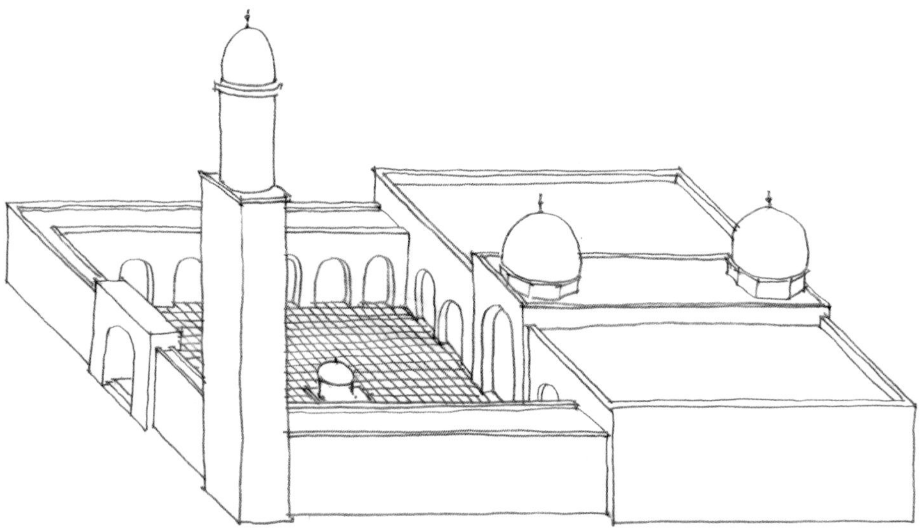

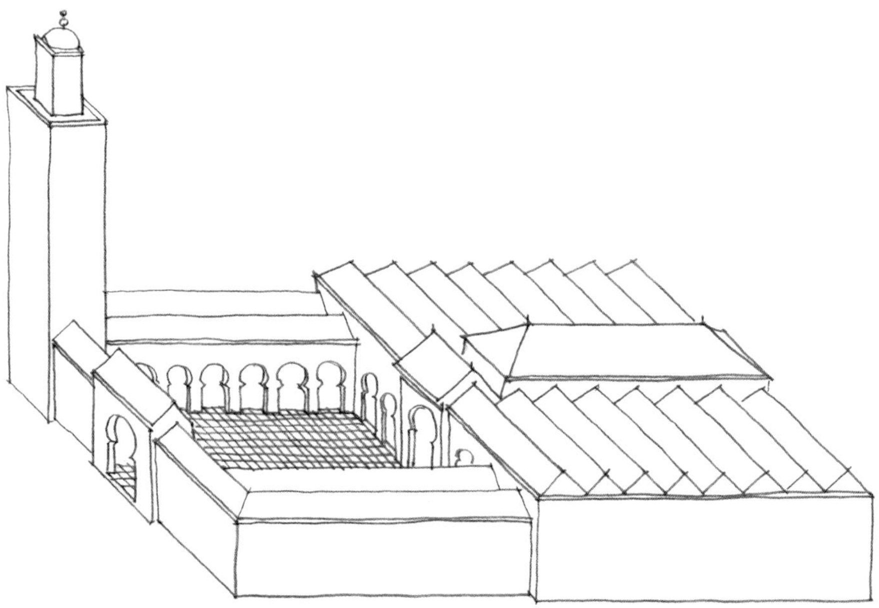

Typical regional architectural styles for mosques

opposite left

Spain and North Africa

opposite right

Anatolia and the Ottoman region

right

Iran and Central Asia

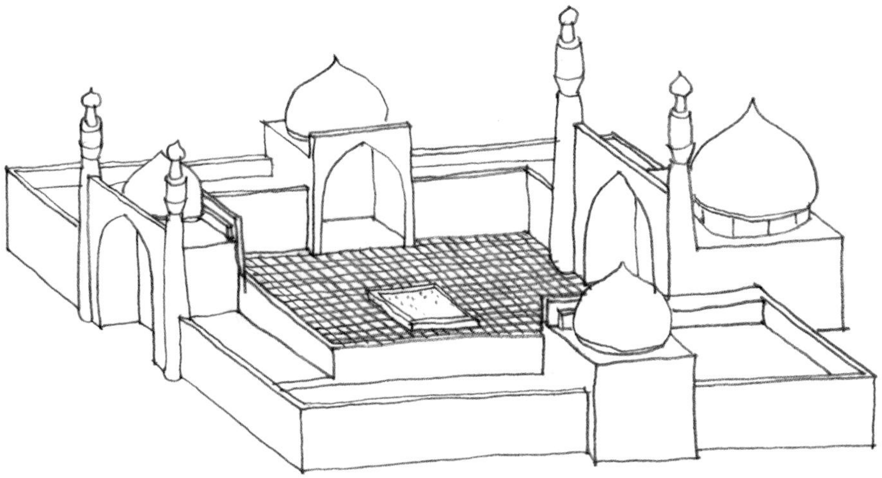

Typical regional architectural styles for mosques

opposite left

West Africa and the Sub-Sahara region

opposite right

The Indian subcontinent

right

South East Asia

far right

East Africa and eastern Arab lands

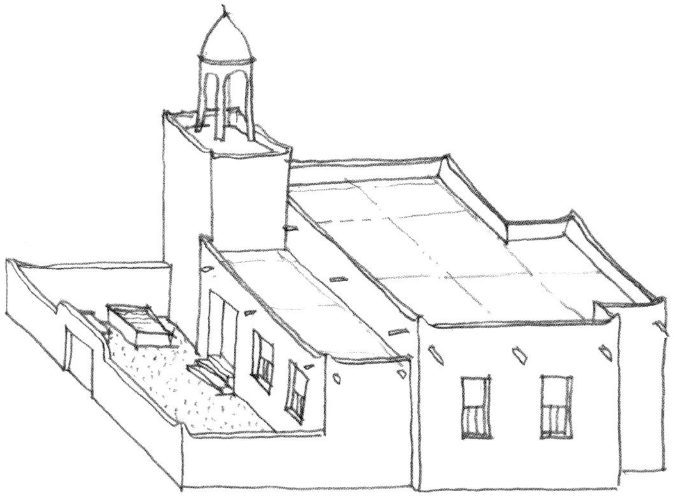

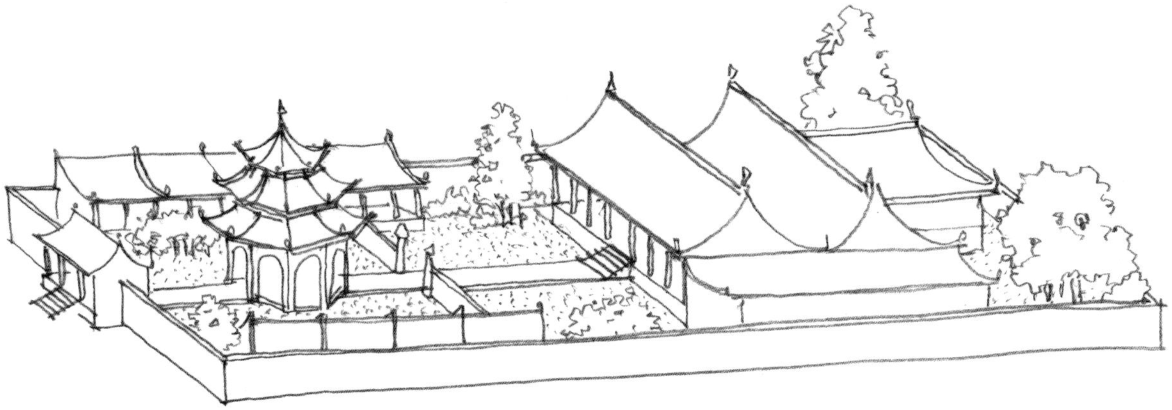

Over time, mosques evolved with a set of typical architectural features. Not all of these features are observed in all of the mosques, but they do not change much due to sectarian differences within Islam. For instance, differences between the Sunni and Shia did not affect the primary architectural features and style of mosques. Their main difference is in the choice of quotations from the Quran used as calligraphic ornament. Today typical architectural features observed in a mosque are as follows:

opposite

Typical regional architectural styles for mosques

The Chinese region

right

Typical architectural features of a mosque

1 Prayer hall or demarcated area for prayer
2 Qiblah wall and the mihrab
3 Mimbar or pulpit in the mosque
4 Dikka or raised tribune in the mosque for Quran and prayer recitals
5 Kursi or stand for placing the Quran in the mosque
6 Minaret
7 Ablution area or pool
8 Portal or main entrance

Many of these characteristic features can be observed in traditional Maldivian mosques together with influences from other regional styles.

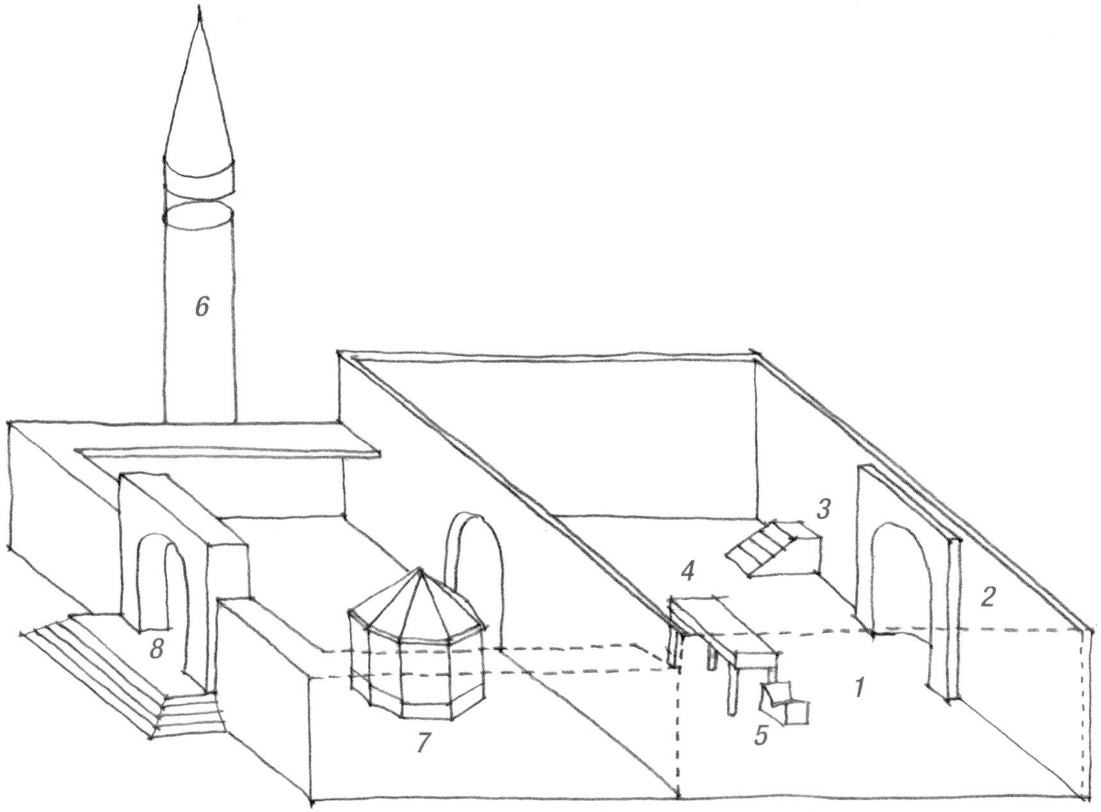

Expression of Cultural Fusion

This fact [that Maldives is a nation of mixed races] is evident even to the casual visitor, when he looks at the faces of the people on a Male' street. The diverse characteristics include those of Arab, Persian, races from the Indian sub-continent, Indonesian, African and Mediterranean ethnicities.

Naseema Mohamed, Maldivian historian, 2008

Coral stone mosques stand as a powerful expression of the fusion of cultures in the Indian Ocean region, their distinctive architecture illustrating the region to an extent not found anywhere else in the world. Natural influences such as the climate, winds, currents and topography are all crucial to the shaping of cultures. However, although the monsoons, the winds and currents, played a significant role in forging Indian Ocean cultures, human interactions and the exchange of ideas contributed the most. Architectural styles evolve similarly and become a historic tapestry of influences due to social contact and exchange of ideas. The tapestry of influences on the architecture of the coral stone mosque of Maldives is derived from the Indian Ocean region along with Maldives own past and maritime cultures. These influences can be traced by comparing the similar architectural features in the mosques and other structures of the Indian Ocean region.

The Indian Ocean region cannot be defined as a common cultural region such as the Mediterranean. Historians like KN Chaudhuri are convinced about the unity of the Indian Ocean in terms of economic and social life, while other historians like Michael Pearson feel the unity is more religion-based, referring to the spread of Islamic culture during the period of Arab trade in the Indian Ocean. Nevertheless, Indian Ocean trade and travel left a trail of cultural spaces with cultural exchanges concentrated among the littoral communities of the ocean rim. Maldives is an example of such a community; it played a crucial role in the travel and trade of the Indian Ocean and was exposed to all the cultures of the region to varying extents throughout its history. The cultural spaces of the Indian Ocean can be defined in four geocultural regions with several subcultural regions. They can be broadly categorised as follows:

1 South Asia
   i) India
   ii) Sri Lanka
   iii) Other coastal regions of South Asia
2 East Africa
   i) The Swahili coast of Africa
3 South East Asia
   i) The Straits of Malacca
4 West Asia
   i) Maritime regions of Arabia and Persia

Although the architectural styles of these cultural regions are different they share many common features. When Islam spread throughout the Indian Ocean region, it did not impose a specific style of architecture. It was integrated into the existing cultures with a subtle non-physical code of conduct defined by the Quran and the Hadith. Different mosque styles attributed to the Indian Ocean region evolved responding to the individual natural and cultural needs of its subcultural regions. They possess many common features of the region and have influenced each other. Coral stone mosques are such an Indian Ocean cultural phenomenon. The following sections are descriptions of the architectural features of structures from different parts of the Indian Ocean region, with characteristics similar to those of the coral stone mosques of Maldives.

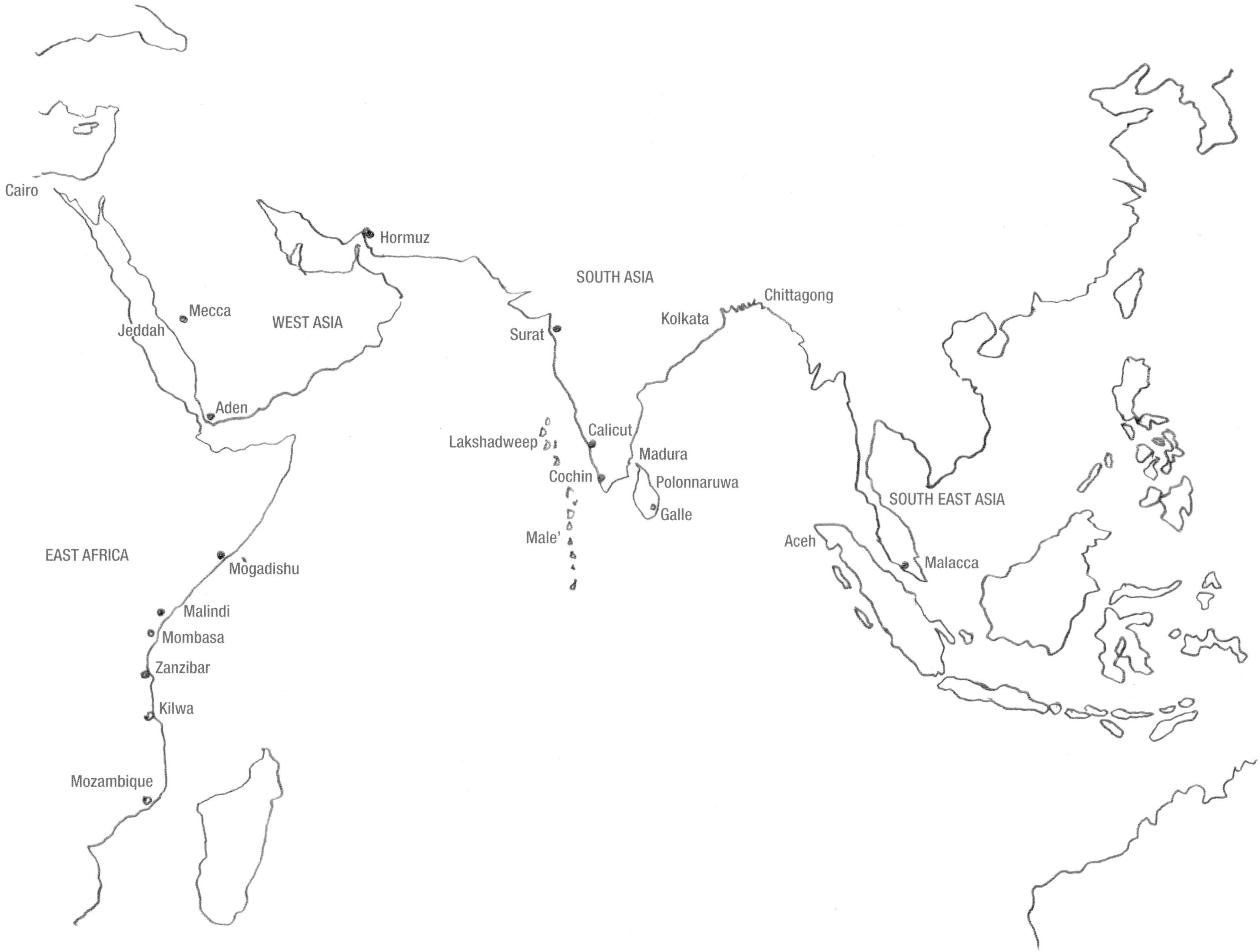
Cairo
Hormuz
SOUTH ASIA
Mecca
Jeddah
WEST ASIA
Chittagong
Kolkata
Surat
Aden
Calicut
Lakshadweep
Madura
Cochin
Polonnaruwa
Galle
SOUTH EAST ASIA
Male'
Aceh
EAST AFRICA
Mogadishu
Malacca
Malindi
Mombasa
Zanzibar
Kilwa
Mozambique

opposite

Geocultural regions of the Indian Ocean

far right

Temple of Lad Khan, Karnataka, India

India

Ancient South Indian architecture uses similar architectural elements to Maldivian coral stone mosques. Indian buildings include elements such as raised plinths, rising stairs, exterior decorative stone mouldings and interior ornamental elements. The Temple of Lad Khan in Karnataka, built in 500 CE, is an example of such a structure where the raised plinths and rising stairs share similarities with the mosque.

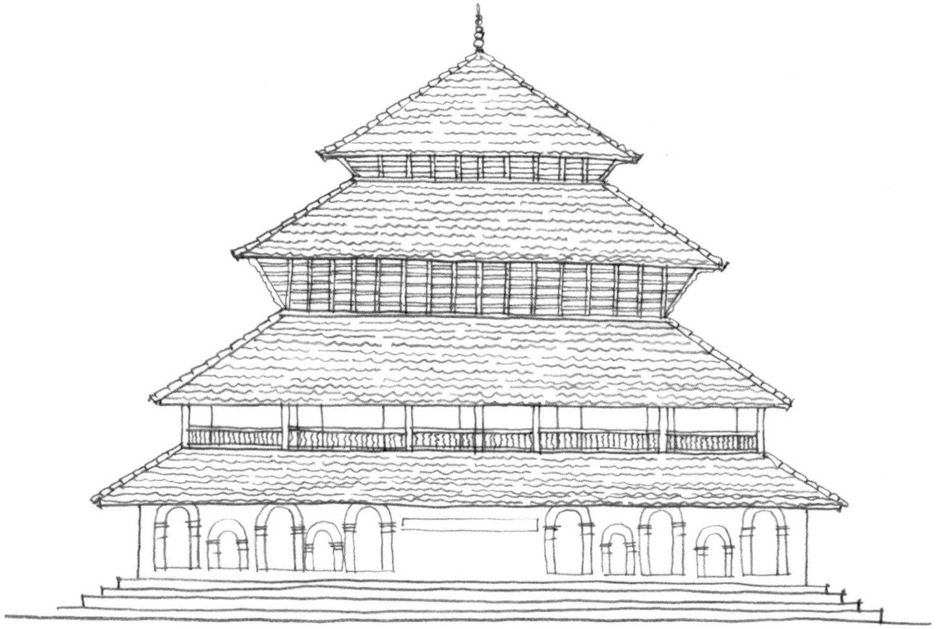

opposite left

Brihadeeswarar Temple, Thanjavur, India

opposite right

Miskhal Masjid, Calicut, India

Brihadeeswarar Temple in Thanjavur, built in 1010, is an example where the exterior decorative elements – such as the rising tapering layers depicting doorways, pilasters and mouldings – are similar to the ornamental concept of the interior *laage'* (a layered recessed part of the ceiling with depictions of tapering layers of doorways, pilasters and mouldings) of the coral stone mosque. These ancient Indian counterparts stand apart from mosque architecture because they used animal and human motifs in their decoration.

South Indian traditional mosques from the Malabar and Coromandel coastal regions have many structures with similar architectural features to Maldivian coral stone mosques. These characteristics include the hypostyle prayer hall, veranda-like antechambers, large overhanging tiered-roof forms and coffered ceilings. Mosques from the Malabar region such as the Jami Masjid and Miskhal Masjid in Calicut, first built around 1400, and Chembitta Palli in Cochin, built around the 1520s, have tiered-roof forms, prayer halls with columns, antechambers and coffered ceilings that can be compared to the coral stone mosques. However, their similarities are not so strong because the timberwork, timber carvings and calligraphy differ so much from their Maldivian counterparts. Mosques in the Coromandel region, such as the Ala al-din Mosque and Shrine in Madura, are examples of similar plan forms and raised decorated plinths, but again, while the types of mouldings are similar, the mosques are of granite and the decorative elements are very different. Keralan vernacular architectural features, such as coffered ceilings and woodcarvings, as found in the Sree Padmanabhapuram Palace near Thiruvananthapuram, also share similarities with the ceilings in Maldivian coral stone mosques.

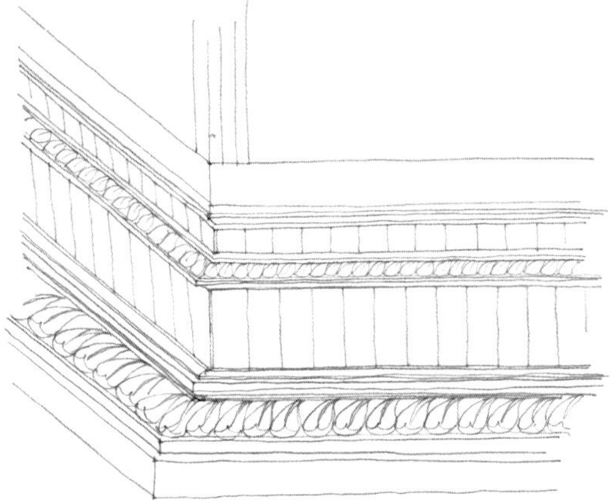

Influences from the Gujarati region of India are more obvious among the ornamental elements of the coral stone mosques of Maldives. The floral carvings and decorative motifs found in the Gujarat have many similarities to the decorative timberwork found in the columns and door partitions of the Maldivian mosques. Evidence of direct interaction between Maldives and the Gujarati region comes from two stone tablets or tombstones from Khambhat in the Gujarati region. One was found in the mihrab wall of Seedhee *Miskiy* in Male' (originally built in around 1335) and the other in Kinolhas, Baa Atoll. The lacquerwork of Kutch and Sankheda, Gujarat, may also have had an influence on the lacquerwork of Maldives.

opposite left

Detail of the central chamber of Ala al-din Mosque and Shrine, Madura, India

opposite right

A 'Khambhat stone' in the mihrab of Seedhee *Miskiy*, Male', Maldives

far right

Ujra Mosque, Karavatti Island, Lakshadweep Islands, India

Mosques on Minicoy and the Lakshadweep Islands of India also share similar features to those of the coral stone mosque, such as the tiered roof and hypostyle prayer hall. Similarities can be observed in Jami Masjid, Minicoy, built around the 1700s and Ujra Mosque, Kavaratti Island in the Lakshadweep Islands, built around the same period. But none of them unites the elements of architectural form, design or the craftsmanship and artistry of the Maldivian coral stone mosques.

Sri Lanka

Ancient Sri Lankan architecture shares similarities with Maldivian coral stone mosques, such as raised plinths, rising stairs and exterior decorative stone mouldings. The monastery complex of the ancient city of Polonnaruwa in Sri Lanka contains examples of structures with raised decorated plinths. HCP Bell traced some of the features and profiles from the raised coral stone platforms of ancient Buddhist structures in Maldives to Sri Lankan Buddhist origins at the Toluvila Monastery in Anuradhapura.

The practice of coral stone construction probably arrived from Sri Lanka and India along with the Buddhist and Hindu cultures. Indirect influences from ancient Sri Lankan Buddhist architecture can be seen in in the pre-Islamic architecture of the Maldivian coral stone structures. The coral stone mosques stand apart from their predecessors, the Maldivian pre-Islamic structures and ancient Indian counterparts, as previously mentioned, have animal or human motifs in the ornament.

The early mosques of Sri Lanka, such as those in the Galle region, had features like hypostyle halls with columns, antechambers, verandas and large overhanging tiered-roof forms similar to many structures of the region and the coral stone mosques of Maldives. However, these early Sri Lankan mosques lacked raised platforms.

Thuparamaya,

Polonnaruwa, Sri Lanka

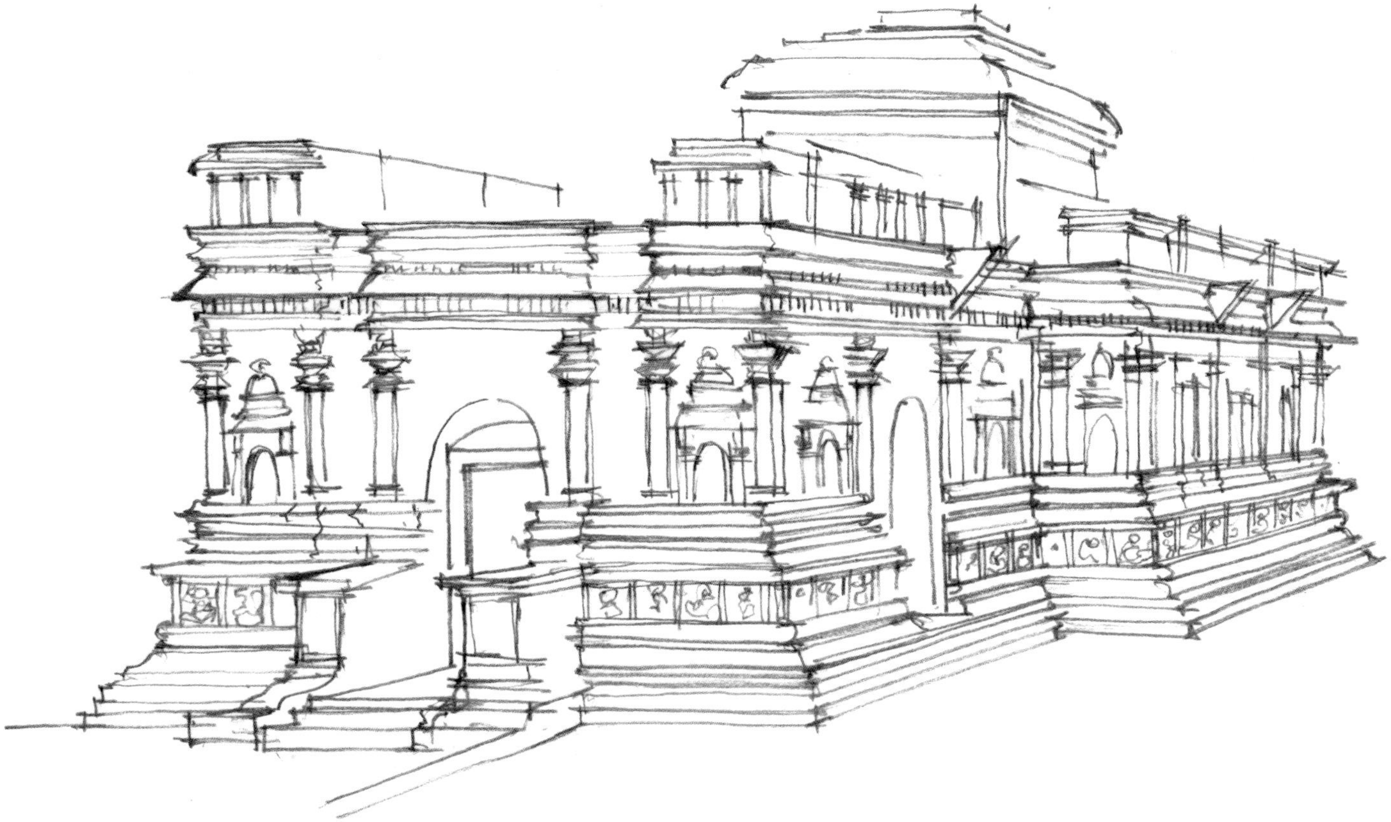

Other coastal regions of South Asia

Maldivians travelled to other coastal regions of South Asia and traded frequently with cities in Bangladesh, Myanmar (Burma) and Thailand, and particularly the port cities of the Bay of Bengal. The obvious influence from this region is the lacquerwork found in the mosque decoration as well as in Maldivian crafts. Lacquerwork, such as the turned lacquered columns, turned pilasters and other lacquer ornament found in the interiors of the coral stone mosques, bear strong resemblance to the lacquerwork of the Myanmar (Burma) region.

The Swahili coast of Africa

Mosques in the Swahili coastal region also have many structures with architectural features similar to those of Maldivian coral stone mosques. These include the hypostyle prayer hall, antechambers, overhanging tiered-roof forms and, most importantly, coral construction. Today some of the finest examples of coral stone architecture can be seen on the Swahili coast area of East Africa. These include

opposite
Lacquerworks from Myanmar (Burma)
right
Early coral tombstone from Kilifi, Kenya

the first stone mosque excavated at Shanga in the Lamu Archipelago of Kenya. The mosque was constructed around the first century from neatly shaped Porite coral bonded with mud, with a white plaster face. Similarities can be observed in the tombstone carvings of the mosque complex at Kilifi, Kenya, where the carvings and iconography have much in common with features of Maldivian coral stone mosques.

A comparable example is also found in the Kizimkazi Mosque in Zanzibar, first built in 1107, where the form and layout are similar but the raised plinth of the coral stone mosque is absent. The coral carvings are found only in the mihrab wall and the construction techniques are different. Coral stone used in structures on the Swahili coast is set in mortar using masonry techniques, while the Maldivian counterparts are shaped to fit and assembled without using mortar.

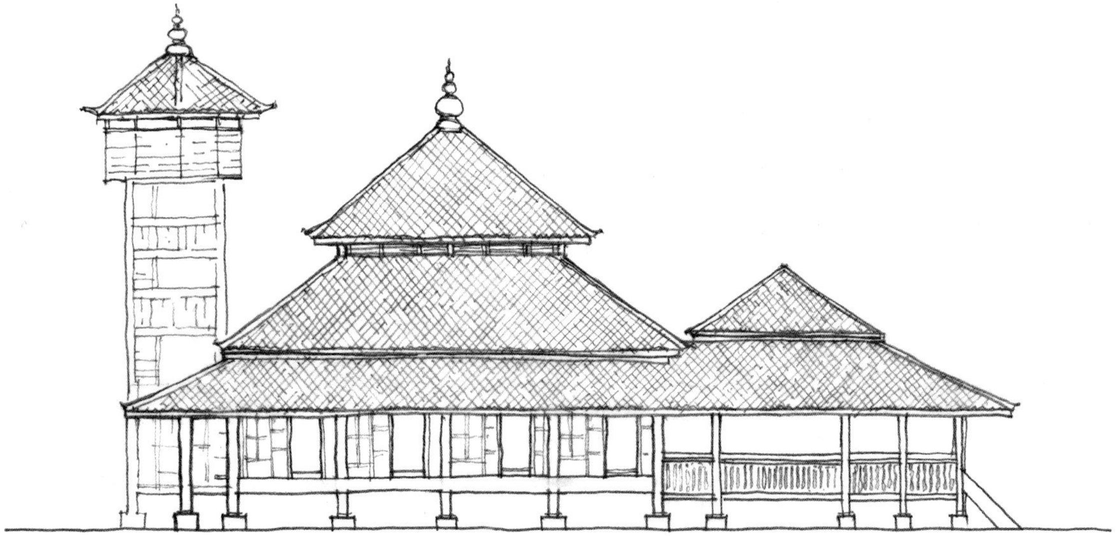

The Straits of Malacca region

Mosques in the Straits of Malacca region have similar architectural features to the Maldivian coral stone mosques, such as the tiered-roof forms, verandas and raised platforms. The raised platforms and most parts of the ancient mosques of this region are made from timber – there are no signs of coral stone or dry stone construction techniques. The roof forms of Masjid Kampung Kling, Malacca, Malaysia, first built in 1748 and Demak Great Mosque in Jakarta, Indonesia, first built in the 1500s are notable examples.

Maritime regions of Arabia and Persia

opposite
Masjid Kampung Laut, Kelantan, Malaysia
far right
Jeli Thuluth style calligraphy from the Mamluk period in the Aslam al-Silahdar complex in Cairo, Egypt

Islam and mosque construction were introduced to Maldives as a direct result of the Arab and Persian traders in the Indian Ocean region. Religious scholars from Yemen, Egypt and Shiraz were major influences on Maldivian culture. The timber latticework, a common feature in the windows of Maldivian coral stone mosques, is comparable to the timberwork of the mashrabiyya. Of course the timberwork of the mashrabiyya is much more sophisticated and elaborate, but close resemblances can be observed in the Male' Old Friday Mosque and Male' Eid Mosque. The Arabic calligraphic decorations used in all the mosques are derived from the ancient Jeli Thuluth style.

Traditional Mosques of Maldives

Their [Maldivians] temples are called Mesquites ... The temple is square, facing west ... They must mount eight or nine steps to the elevation of the temple. The paved floor of the temple is covered with pretty mats and carpets ... The superstructure is of wood, the carpentry of which I admire much, for which it could not be better polished or worked. The walls are wainscoted with wood, worked and fitted in the same way; and the whole of the woodwork, outside and in, is put together without nail or bolt of any kind, and yet holds so fast that one could not take it to pieces unless one knew the artifice. You see large slabs, either of stone or timber ... which are engraved letters and inscriptions in the Arabic language ... throughout the temple, which is spacious ... The temples, or mesquites, are very numerous in all the inhabited islands ...

Francois Pyrard, French navigator and adventurer, 1619

The history of Maldives goes back a few thousand years, but today the oldest buildings in the country are the surviving traditional mosques dating from the 1300s. They are the most powerful visible expressions of the islands' rich cultural heritage, which is eroding at an alarming rate. Most of the traditional structures in Maldives do not survive long because they are not made from durable materials, however, most of the mosques are lost due to neglect or to make way for new developments. In 1940, the former archaeological surveyor of the British Government, HCP Bell, identified 32 traditional mosques on the capital island Male', yet today only six of them remain.

The earliest Maldivian mosques, built soon after the conversion to Islam in 1153, did not survive in their original state as they were rebuilt and remodelled many times. Many were built on the sites of temples or monasteries from the pre-Islamic period; building materials and traditions of the period were reused and reinterpreted to suit the requirements of the new buildings, as seen in Kudahuvadhoo Old Friday Mosque and Nilandhoo Old Friday Mosque, where both sites were former Buddhist places of worship.

Types of Traditional Mosque

Today four types of traditional mosque survive with their unique architectural features. These bear witness to an interesting historic tapestry of cross-cultural exchanges, the maritime culture and vernacular architecture. The four types were built between the 14th and the 20th century and are categorised by their use of material and type of construction. They are as follows:

1 *Mosques using* veligaa *and shaped stone masonry construction techniques.* The plinth and walls are built from shaped *veligaa* blocks set in mortar and finished with a lime skim coat. The plinth is low with simple mouldings and no ornament. The roof and ceiling structure are simpler than in the other types of mosque, they do not have lacquered interiors and a *laage'* (stepped, recessed ceiling). These are the earliest type of mosque and are few in number. An example is Gen *Miskiy* in Fuvah Mulah; the date of its construction is unclear, but it was certainly built earlier than 1397 and was probably one of the first mosques built after the island converted to Islam.

2 *Mosques using both* veligaa *and* hirigaa *and shaped stone masonry construction techniques.* The plinth and walls are built from both *veligaa* and *hirigaa* blocks set in mortar and finished with a lime skim coat. The platform bases have mouldings but do not have intricate ornamental work. The roof and ceiling structure are decorated with a *laage'* and lacquerwork. Most of these mosques are probably rebuilt or restored old *veligaa* mosques where part of the old structure was used. A typical example of such a mosque is Vaadhoo Old Friday Mosque, believed to be built in the late 1500s.

3 *Mosques using* hirigaa *and shaped stone carpentry techniques.* The plinth and walls of these mosques are built using smooth and highly decorated interlocking blocks of *hirigaa* coral stone assembled without mortar or plaster. They have detailed mouldings and intricate carvings. The roof and ceiling structure have coffered ceilings with a *laage'* and intricate timber and lacquer-decorated interiors. These are the best examples of mosques and are widely distributed throughout the country. They display the skills and craftsmanship of people from a bygone age. Examples of this type of mosque are Male' Old Friday Mosque built in 1657, Male' Eid Mosque built in 1815 and Fenfushi Old Friday Mosque built in 1695.

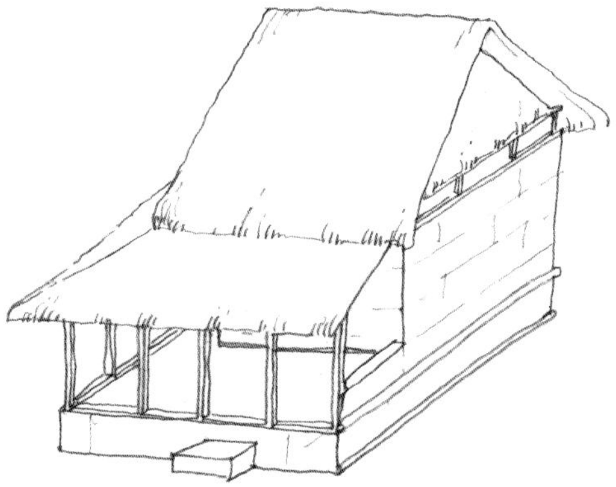

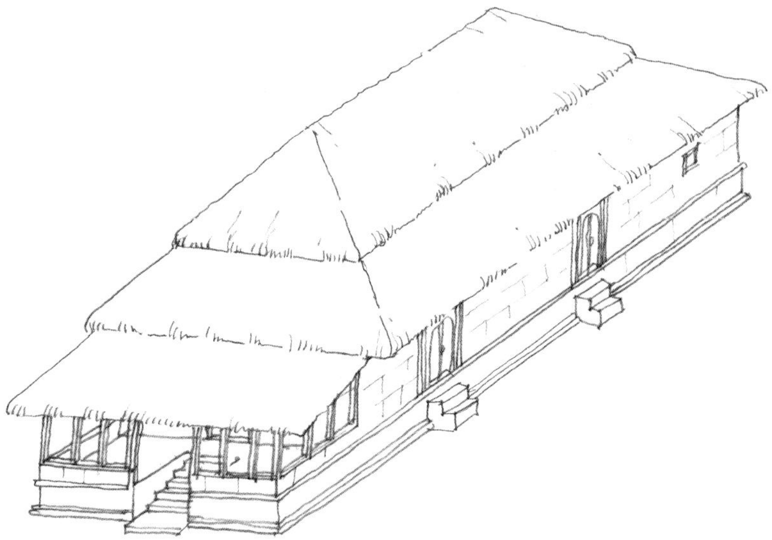

4 *Mosques using* thelhigaa *or chipped coral stone and masonry for construction*. The plinth and walls of these mosques have smoothly plastered coral masonry using chipped coral stone, coral lime and sand mortar with smooth lime plaster. The plinth is lower with few mouldings and no decorative carvings. These mosques were introduced during the early 20th century when masonry became popular. The roof and ceiling structure is simpler than this type's predecessor. There are many such mosques but most of them are still not recognised as cultural assets and are very much endangered. One such example is Male' Bihurozukamana Mosque, rebuilt in 1919.

opposite left

Gen *Miskiy*, Fuvah Mulah, Gnaviyani Atoll

opposite right

Vaadhoo Old Friday Mosque, Vaadhoo, Gaaf Dhaal Atoll

right

Male' Old Friday Mosque, Male'

far right

Male' Bihurozukamana Mosque, Male'

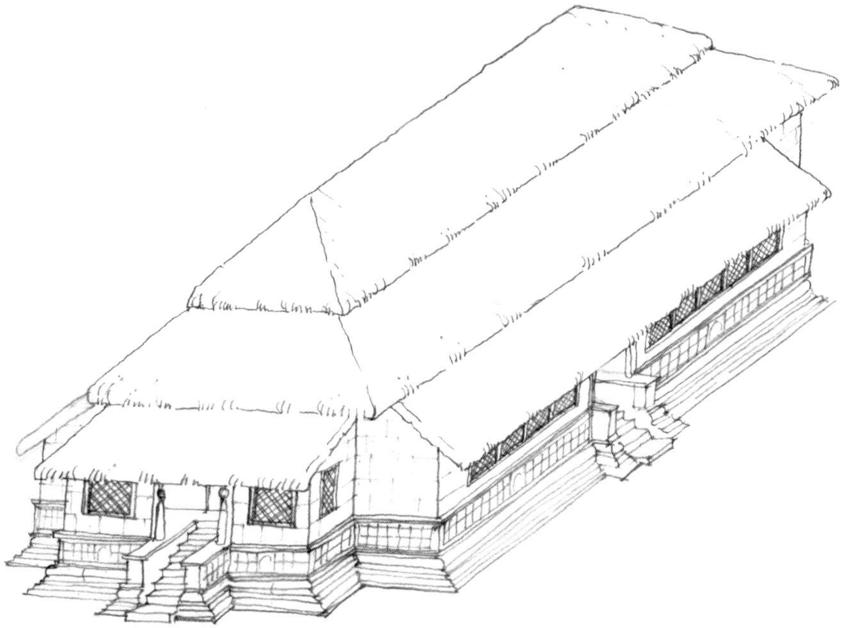

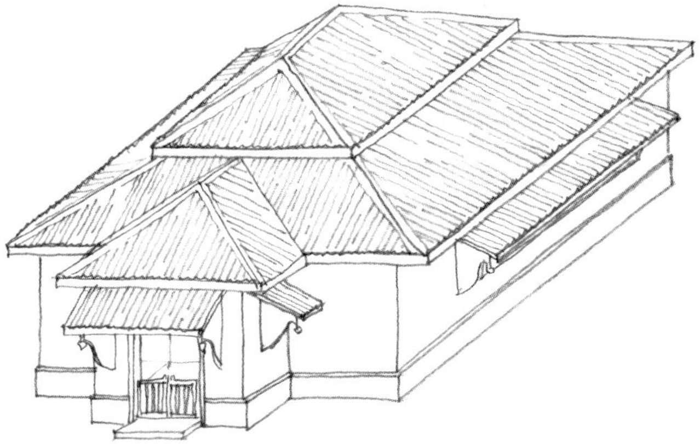

The Traditional Mosque Compound

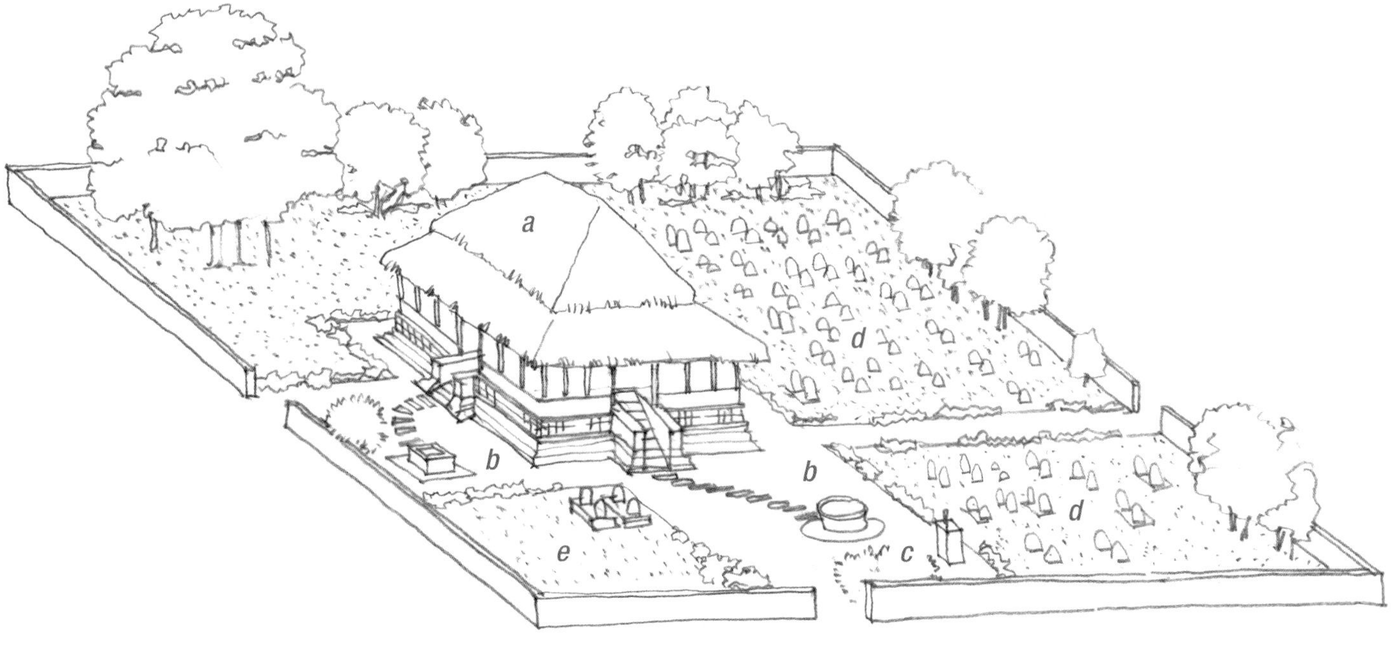

A typical traditional mosque compound:

a mosque building

b water wells

c sundial

d cemetery

e mausoleum

The traditional mosque is a simple building within a mosque compound. Features of the mosque compound are typical and include water wells, a sundial, cemetery and occasionally small minarets and bathing tanks. The wells located near the entrances of the mosque are used for ablutions and as a water supply for the community. Stepping stones from the well lead up to the mosque so that feet can be kept clean on entering. The wells are about a metre across and three metres deep and water is drawn using a *dhaani* – a container attached to a timber pole. The sundial showing prayer times and the qiblah is in a clear area near the entrance of the mosque compound and is mounted on a pediment.

The cemetery is lined with tombs and mausoleums. Each tomb has two tombstones; the tombs of males have a pointed design, while those of females are rounded. The tombstones are made from intricately carved one-piece coral stones. Mausoleums are tombs of revered people so are placed within an enclosed shelter. Boundary walls and entrance gates are occasionally found if a mosque is located in a populated area. The landscaping with paths leading to the mosque and wells is neatly laid with white sand. Plants include jasmine and mint near the mosque wells and frangipani trees mark the mosque compound. Holy basil, lilies and flowering trees are also planted in the cemetery area.

Traditional Mosque Features

Typical features of a traditional mosque:

a prayer hall

b *dhaala*

c mihrab wall

d rising stairs

e raised plinth

f tiered roof form

g *laage'* and coffered ceiling

Typical architectural features include a rectangular prayer hall and a combination of the veranda-like antechambers called *fenda* or *dhaala* and occasional mihrab chambers. The mosque is built on a raised plinth and entered through rising stairs from the eastern side. On both sides of the stairs are timber ledges that are used as seating for the elderly and the disabled. The roof is tiered, hipped and steeply sloped with large overhangs and is finished with either plaited coconut thatching, clay roofing tiles or metal roofing sheets. All mosques have a timber ceiling and many have a central stepped recessed area called a *laage'*. The doors vary depending on the type of traditional mosque but most have sliding doors. The iconography of decoration is based on organic and geometric patterns with calligraphic writings from the Quran in Thuluth Arabic script.

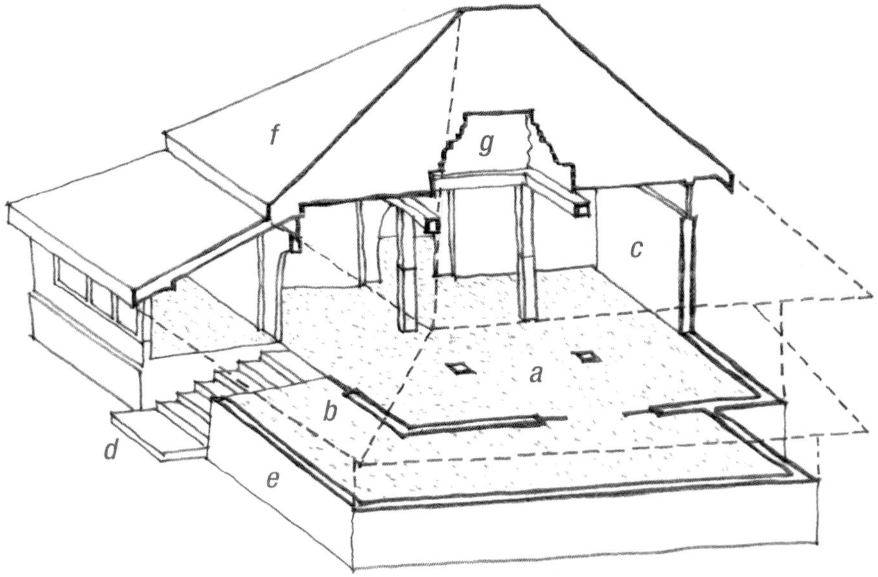

Traditional Mosques Noted by the Authors

Veligaa (Coral Sandstone) Mosques: 9

The following is not a comprehensive list of all the traditional mosques of Maldives but is limited to those noted by the authors.

Fuvah Mulah Gen Mosque – *c*1300

Gen *Miskiy*, Fuvah Mulah (Gnaviyani Atoll), is known as the oldest mosque on the Maldivian island and was built straight after its conversion to Islam. The date of construction is not clear but it is on the site of a Buddhist monastery and predates Kedeyre *Miskiy* built *c*1397. The mosque compound has a well, a cemetery with tombs, and a bathing tank. The mosque has a simple prayer hall with small open verandas and no decoration or carvings. It is constructed from *veligaa* blocks and timber.

Fuvah Mulah Kedeyre Mosque – *c*1397

Kedeyre *Miskiy*, Fuvah Mulah (Gnaviyani Atoll), was probably built around the same period as Koagannu Mosque, Hulhumeedhoo (Seenu Atoll), *c*1397. The mosque compound has a well, a cemetery with tombs, and a bathing tank. The mosque building has a simple prayer hall with an open veranda at the entrance. There are no decorations or carvings and the mosque is constructed from *veligaa* blocks and timber.

Hulhumeedhoo Koagannu Mosque – *c*1397

Koagannu *Miskiy*, Koagannu, Hulhumeedhoo Island (Addu City), is known as the oldest mosque in Koagannu. Koagannu is the largest and the oldest cemetery compound in the country with many mosques, wells, tombs and mausoleums. This mosque was probably built around 1397 but has been extensively restored to the extent that its originality has been lost. The mosque has a simple prayer hall with small open verandas and no decoration or carvings. It is constructed from *veligaa* blocks and timber.

Hulhumeedhoo Athara Mosque – *c*1417

Athara *Miskiy*, Koagannu, Hulhumeedhoo Island (Addu City), is known as the third oldest mosque in Koagannu. Like Koagannu mosque, this mosque is also situated in the Koagannu cemetery compound. The mosque was probably built around 1417 using *veligaa* sandstone block construction and it has no timber decoration or carvings. It has been extensively restored to the extent that its originality has been lost. The mosque has a simple prayer hall with small open verandas and no interior decoration or carvings.

Kolhufushi Gazee Mosque – 1573

Gazee *Miskiy*, Kolhufushi Island (Meemu Atoll), was probably first built around 1573, and is believed to be constructed from the timber of *Kalhuohfummi*, the famous boat of the Utheemu brothers. The islanders later rebuilt the mosque after a fire burned down part of it. During an inspection of damage after the 2004 tsunami, it was confirmed that partly charred timber from an old boat (probably *Kalhuohfummi*) still remains. It is a simple building with most of the features of a traditional mosque. The exterior ornament is simple with mouldings on the plinth while the interior decoration is intricate, with timber carvings, lacquerwork and calligraphy. The mosque is constructed from *veligaa* blocks and timber.

Hangnaameedhoo Old Mosque – *c*1600

The Old Mosque, Hangnaameedhoo (Alif Dhaalu Atoll), is near the mausoleum of Kalaafaanu. Kalaafaanu or Sultan Ibrahim III – son of Sultan Ghazi Mohamed – was martyred in a sea battle against Malabari raiders in 1609. The mosque is constructed from *veligaa* blocks and timber and has recently been restored.

Mathiveri Old Mosque – *c*1687

The Old Mosque, Mathiveri (Alif Alif Atoll), was built by Sultan Ibrahim Iskandhar I (1648–1687). It is a small mosque constructed from *veligaa* blocks and timber. It has recently been restored.

Maaenboodhoo Old Friday Mosque – *c*1700

The Old Friday Mosque, Maaenboodhoo (Dhaal Atoll), is built on the site of a Buddhist monastery and it is not clear if there had been an earlier mosque. The present mosque was first built probably around the 1700s but was extensively restored in recent times. The mosque has fine interiors with lacquerwork and is constructed from *veligaa* blocks and timber.

Maaenboodhoo Small Mosque – *c*1700

The Small Mosque, Maaenboodhoo Island (Dhaal Atoll), was probably built around the 1700s. The mosque has been restored extensively but still has fine interiors with lacquerwork. It is built using *veligaa* blocks and timber.

Veligaa/Hirigaa (Coral Sandstone and Porite Coral) Mosques: 7

Vaadhoo Old Friday Mosque – *c*1500

The Old Friday Mosque, Vaadhoo Island (Gaaf Dhaal Atoll), was probably built around the 1500s as it was known to have been used by the religious scholar Vaadhoo Dhanna Kaleyfaanu. The mosque compound also has his mausoleum. It was extensively restored by Sultan Mohamed of Dhevvadhoo (1692–1701), and restored again in 1989. The mosque compound has the mosque, a stone well, sundial and cemetery with mausoleums and tombstones. It is the largest traditional mosque outside Male' and has two prayer halls, mihrab chamber, two antechambers next to the mihrab chamber, a *dhaala* on one side, rising main entrance stairs and four other entrances. It is built on a plinth with walls, columns, a two-tiered roof, a coffered ceiling with five *laage'*, sliding doors and windows. The exterior ornament has mouldings on the plinth and inside there are timber carvings, lacquerwork and calligraphy. The mosque is constructed from both *veligaa* and *hirigaa* blocks and timber.

Utheemu Kandhuvalu Mosque – *c*1500

Kandhuvalu *Miskiy*, Utheemu (Haa Alif Atoll), was probably first built in the 1500s as it has the mausoleum of the father and the grandfather of Sultan Ghazi Mohamed Thakurufaanu (1573–1585). It was rebuilt in 1878, restored in 1946 and in the 1990s and remains in use as a mosque today. The mosque compound has the mosque, a well and a cemetery with mausoleums. The mosque has a prayer hall, mihrab chamber, *dhaala* on three sides and rising entrance stairs. It has a simple plinth, walls, columns, a two-tiered roof, a coffered ceiling with l*aage'* raised using *thona*, sliding doors and windows with *miskithu jaali*. The exterior decoration is simple with mouldings on the plinth while the interior ornament consists of timber carvings, lacquerwork and calligraphy. The mosque is constructed from both *veligaa* and *hirigaa* blocks and timber.

Baarah Old Friday Mosque – *c*1500

The Old Friday Mosque, Baarah (Haa Alif Atoll), was probably first built in the late 1500s, and used by Sultan Ghazi Mohamed Thakurufaanu (1573–1585). It was later restored by Sultan Mohamed of Dhevvadhoo (1692–1701) and last restored in 2002. The mosque compound has the mosque, a well, a small minaret and a cemetery. The mosque has a prayer hall, mihrab chamber, *dhaala* on three sides and rising entrance stairs. It has a simple plinth, walls, columns, a two-tiered roof, a coffered ceiling with *laage'* raised using *thona*, sliding doors and windows with *miskithu jaali*. The exterior ornament is simple with mouldings on the plinth. The interior decoration is intricate with timber carvings, lacquerwork and calligraphy. The mosque is constructed from both *veligaa* and *hirigaa* blocks.

Male' Dharumavantha Rasgefaanu Mosque – 1694

Dharumavantha Rasgefaanu *Miskiy*, Male', is believed to have been originally built by the first Muslim Sultan, Mohamed Ul-Adil, known as Dharumavantha Rasgefaanu (1153–1163). In 1694 it was rebuilt, extensively restored in 1931 and conserved using modern techniques in 2005. The mosque has a hypostyle prayer hall with columns, and rising entrance stairs with *dhaala* at the entrance. It has a simple plinth, walls, columns, a two-tiered roof, a coffered ceiling with *laage'* raised using *thona*, sliding doors and windows with *miskithu jaali*. The exterior ornament is simple with mouldings on the plinth. The interior decoration is intricate with timber carvings, lacquerwork and calligraphy. The mosque is constructed from both *veligaa* and *hirigaa* blocks and timber.

Male' Seedhee Mosque or Henveiru Old Mosque – 1697

Seedhee *Miskiy* or Henveiru Old Mosque, Male', is believed to have been built originally by Sultan Jamaludeen Umar Veeru (1306–1335). It was rebuilt in 1697 by Sultan Mohamed of Dhevvadhoo and remodelled extensively using masonry in 1919. The mosque is a small building with a prayer hall, *dhaala* on three sides and rising entrance stairs. It has a simple plinth, walls, columns, a two-tiered roof, a coffered ceiling with *laage'*, doors and windows with *miskithu jaali*. The exterior decorations are simple with mouldings on the plinth. The interior decoration has no carvings, lacquerwork or calligraphy. The mihrab has a *c*14th-century marble stone from Khambhat, Gujarat. The mosque is constructed from *hirigaa* blocks coral rubble and timber.

Dhevvadhoo Old Friday Mosque – *c*1701

The Old Friday Mosque, Dhevvadhoo (Gaaf Alifu Atoll), was first built by Sultan Ibrahim III (Kalaafaanu) (1585–1609) and rebuilt in 1700 by Sultan Mohamed of Dhevvadhoo (1692–1701). It was extensively renovated using masonry and metal roofing in the 1920s and again in 2001, but it still has decorated interiors. It is constructed from *veligaa* and *hirigaa* blocks.

Naalaafushi Old Mosque – *c*1700

The Old Mosque, Naalaafushi (Meemu Atoll), is partly constructed of coral stone, although the date of construction is unclear. It was damaged during the 2004 tsunami but has since been renovated.

Hirigaa (Porite Coral) Mosques: 20

Guraidhoo Ziyaaraiy Mosque – *c*1400

Ziyaaraiy *Miskiy*, (Thaa Atoll), was probably built in the 1400s. The mosque complex has the mosque and the mausoleum of Sultan Usman of Fehendhoo who was banished to Guraidhoo in 1388. The mosque has a small prayer hall, *dhaala* on three sides, one entrance stair, a two-tiered roof, a simple coral base and a coffered ceiling with *laage'*. It was built using coral stone or *hirigaa* along with timber-frame construction, but over time its integrity has been compromised by extensive remodelling. The mosque does not have fine coral carvings but the timberwork is of a fine standard.

Hulhumeedhoo Boadha Mosque – *c*1403

Boadha *Miskiy* in the Koagannu area of the island of Hulhumeedhoo (Addu City), was probably built around 1403 and continues to be used as a mosque today. The Koagannu area has the largest and oldest cemetery in Maldives with many mosques, wells, tombs and mausoleums. It has four small mosques including this one. Boadha *Miskiy* has a small prayer hall linked to an open veranda at the entrance, built on a small platform. The mosque is constructed from *hirigaa* blocks and timber but it has been renovated to such an extent that the original materials are hardly visible.

Hulhumeedhoo Fandiyaru Mosque – *c*1586

Fandiyaru *Miskiy* in the Koagannu area of the island of Hulhumeedhoo (Addu City), was probably built around 1586 during the reign of Sultan Ibrahim III and continues to be used as a mosque today. Koagannu has the largest and oldest cemetery in Maldives with more than 500 coral stone tombs and 15 mausoleums. One of four small mosques, this one has a little single-space prayer hall, an open veranda at the entrance linked to a well by stepping stones and a small minaret. It is built on a small platform of coral with mouldings and has the features of a traditional mosque. The mosque is very basic and has no exterior or interior ornament. It is constructed from *hirigaa* blocks and timber.

Veyvah Old Mosque – *c*1600

The Old Mosque, Veyvah (Meemu Atoll), was probably built approximately 400 years ago and continues to be used as a mosque today. It is a small coral stone mosque, with *dhaala* as later extensions, and does not have fine ornamentation in terms of coral carving and lacquerwork. All the structures are in a poor dilapidated condition. The mosque has a prayer hall, *dhaala* on three sides, one entrance stair, a two-tiered roof, a simple coral base and a coffered ceiling. The mosque complex does not have boundary walls but has two wells, a sundial and a cemetery with coral tombstones.

Bileddhoo Old Mosque – *c*1600

The Old Mosque, named 'Masjid Zikra', on Bileddhoo (Faaf Atoll), has an unclear construction date but it is an old coral stone mosque that continues to be used today. The mosque has been remodelled extensively compromising its integrity, and does not have fine ornamentation in terms of coral carvings and lacquerwork. The building has a prayer hall, mihrab chamber, *dhaala* on three sides, one entrance stair, a two-tiered roof, a simple coral base and a coffered ceiling. The mosque complex has the mosque, a coral stone well and a few large coral tombstones.

Male' Old Friday Mosque – 1658

The Old Friday Mosque on the capital island of Male', was built in 1658, replacing the original mosque built in 1153 by the first Muslim Sultan of Maldives. It is the largest coral stone mosque in Maldives and one of the finest coral stone buildings in the world. The mosque has two main prayer halls, a mihrab chamber, *dhaala* on three sides, three entrance stairs, a three-tiered roof, a coral base with carvings and mouldings and coffered ceilings with three main *laage'*. The mosque compound has the mosque, a large minaret, three coral stone wells, a sundial and a large cemetery with mausoleums and tombs of past kings and dignitaries of the country.

Kondey Old Mosque – *c*1687

The Old Mosque on the island of Kondey (Gaaf Alifu Atoll), is believed to have been first built by the two master carpenters (Ali Maavadi Kaleyfaanu and Mahmud Maavadi Kaleyfaanu) of Male' Hakuru *Miskiy* using its leftover coral stone and timber. It was probably built during the reign of Sultan Ibrahim Iskandhar I (1648–1687). The present mosque was rebuilt and recently was extensively remodelled. Now it is basically a modern structure in traditional mosque form. The cemetery has been damaged and there are hardly any coral tombstones standing. The coral tombstones lying on the compound display high-quality carvings. The mosque has a simple prayer hall with an open veranda at the entrance. There are no carvings or decoration and the mosque is constructed from masonry and cement plastering and timber.

Nolhivaram Old Friday Mosque – 1699

The Old Friday Mosque of Nolhivaram (Haa Dhaal Atoll), was built in 1699 during the reign of Sultan Mohamed of Dhevvadhu (1692–1701) and continues to be used as a mosque today. It is an example of a small coral stone mosque with *dhaala*, but over time extensive remodelling has compromised its originality. The mosque has a prayer hall, mihrab chamber, *dhaala* on three sides, rising entrance stair, a two-tiered roof, a coral base with mouldings and coffered ceilings.

Fonadhoo Old Friday Mosque – *c*1700

The Old Friday Mosque, Fonadhoo (Laamu Atoll), has an unclear time of construction but it is an old coral stone mosque that continues to be used as a mosque today. It has a simple prayer hall, *dhaala* and mihrab chamber.
The mosque compound contains the mosque, one coral stone well and a few coral tombstones. The mosque has been extensively remodelled, compromising its integrity, and it does not have fine ornamentation in terms of coral carvings. Although there is evidence of decoration in the timberwork, there is no lacquerwork.

Ihavandhoo Old Friday Mosque – 1701

The Old Friday Mosque, Ihavandhoo (Haa Alif Atoll), was completed on 16 December 1701 and is still used as a mosque today. It is one of the finest surviving examples of a small coral stone mosque but over time major extensions have compromised its visual integrity. The mosque has a prayer hall, rising entrance stair, *dhaala* on three sides, a three-tiered roof, a coral base with carvings and coffered ceilings with *laage'*. The mosque compound includes a well, cemetery and a small minaret. The cemetery has tombstones of eminent people with rare, high-quality carvings.

Kelaa Old Friday Mosque – 1701

The Old Friday Mosque, Kelaa (Haa Alif Atoll), was built between 1692 and 1701 and continues to be used as a mosque today. It is a small coral stone mosque with *dhaala* but, over time, extensive remodelling has compromised its originality. The minaret is in ruins but the site has coral stone tombstones of exceptional quality. The mosque has a prayer hall, rising entrance stair, *dhaala* on three sides, a single-tiered roof, a coral base with mouldings and coffered ceilings with *laage'*. The mosque compound includes two wells and a cemetery with tombstones.

Fenfushi Old Friday Mosque – 1701

The Old Friday Mosque of Fenfushi Island (Alif Dhaalu Atoll), was built during the reign of Sultan Mohamed of Dhevvadhu (1692–1701) on the site of an earlier mosque and is still used as a mosque today. It is one of the finest examples of a coral stone mosque with all its elements in good condition. The mosque has a prayer hall, mihrab chamber, three entrance stairs, a two-tiered roof, a coral base with carvings and mouldings, and coffered ceilings with *laage'*. The mosque compound has the mosque, a unique coral stone bathing tank, three coral stone wells, a sundial and a large cemetery with tombstones of fine quality.

Isdhoo Old Mosque – 1701

The Old Mosque, Isdhoo (Laamu Atoll), was built in 1701 during the reign of Sultan Ali VII and is still in use as a mosque today. This is the mosque where the royal copper chronicles *Isdhoo Loamaafaanu* were kept. It is one of the finest surviving examples of a small coral stone mosque with *dhaala* and has been maintained with minimal changes and in good condition. The mosque has a prayer hall, mihrab chamber, *dhaala* on three sides, one entrance stair, a two-tiered roof, a simple coral base and a coffered ceiling with *laage'*. The mosque compound has the mosque, an old well and a cemetery with tombstones. This mosque is situated on a remote part of the island, there are no boundary walls.

Nilandhoo Old Friday Mosque – *c*1701

The Old Friday Mosque, Nilandhoo (Faaf Atoll), was probably erected by the first Muslim Sultan of Maldives (1153–1165) and was built on the site of a Buddhist monastery. The mosque compound has the mosque, an old gateway from the Buddhist period, two coral stone wells, and a cemetery with high-quality coral tombstones. The mosque has been extensively remodelled, compromising its originality, but it is still in use today. It is an example of a coral stone mosque with *dhaala* and mihrab chamber. The mosque does not have fine ornamentation in terms of coral carving and lacquerwork.

Kudahuvadhoo Old Friday Mosque – *c*1701

The Old Friday Mosque, Kudahuvadhoo (Dhaal Atoll), was built during the reign of Sultan Mohamed of Dhevvadhoo (1692–1701), replacing the original mosque built by the first Muslim Sultan of Maldives (1153–1165). The mosque compound has the mosque, an old gateway from the Buddhist period, two coral stone wells, and a cemetery with high-quality coral tombstones. The mosque has been extensively remodelled compromising its originality but continues in use today. It is an example of a coral stone mosque with *dhaala* and mihrab chamber. The mosque does not have fine ornamentation in terms of coral carving.

Bandidhoo Old Mosque – c1701

The Old Mosque, Bandidhoo (Dhaal Atoll), was probably built in the 1700s. It is a small mosque with a prayer hall, mihrab chamber, *dhaala* on three sides, one entrance stair, a two-tiered roof, a simple coral base and a coffered ceiling with *laage'*. The mosque does not have fine coral carvings, but the timberwork is of a fine standard. The mosque was probably restored in the 1940s when a masonry wall with windows using *miskithu jaali* and a metal roof were added.

Meedhoo Old Friday Mosque – 1705

The Old Friday Mosque, Meedhoo (Raa Atoll), was probably built around 1705 and is still used as a mosque today. It is a coral stone mosque with *dhaala* and mihrab chamber. The coral workmanship and interior calligraphy is not of a high quality, but the mosque is well maintained with few changes having been undertaken since the 1950s. It is also the only mosque with Indian clay roofing tiles from the 1950s. The cemetery has a collection of coral stone tombs with a high standard of carvings and workmanship. The mosque has a prayer hall, mihrab chamber, *dhaala* on three sides, rising entrance stair, a two-tiered roof, a coral base with mouldings and coffered ceiling with *laage'*. The compound includes three wells, cemetery with tombstones and mausoleum and a sundial.

Male' Kalhuvakaru Mosque – 1789

Kalhuvakaru *Miskiy*, Male', was built in 1789 and is still in use as a mosque today. It is a small coral stone structure with *dhaala* and high-quality workmanship in coral carving and detailed lacquerwork. What makes the mosque exceptional is the use of ebony for its construction. The mosque and the coral stone well have been relocated twice. The mosque has a small prayer hall, *dhaala* on three sides, rising entrance stair, a two-tiered roof, a coral base with carvings and mouldings and a coffered ceiling with *laage*'. Near to the mosque there is a coral stone well.

Thelhigaa (Chipped Coral Stone) Mosques: 7

Male' Eid Mosque – 1815

The Eid Mosque, Male', was built in 1815 replacing an older mosque built during the reign of Sultan Imaduddin (1620–1648) and it is still in use as a mosque today. It is the last of the coral stone mosques and has the best ornamentation and craftsmanship of all the mosques in the country. Over time, the mosque compound has been reduced drastically and extensions have been added to the old mosque, but it is well preserved. The building has a prayer hall, *dhaala* on three sides, one rising entrance stair, a two-tiered roof, a coral base with carvings and mouldings and coffered ceilings with *laage'*.

Huraa Old Mosque – *c*1925

The Old Mosque, Huraa (Kaafu Atoll), was probably first built by Sultan Ghazi Hassan Izzuddin or Dhonbandaara (1759–1767) as it was attributed to him in a scroll found in the mosque. The present mosque was built in 1925 and later restored in 1994. The mosque compound has the mosque, an eight-sided coral stone well and a cemetery with two mausoleums and coral tombstones. The mosque has a prayer hall, mihrab niche, a front *dhaala* and rising stairs. It has a simple plinth, walls, columns, a two-tiered roof, timber ceiling, and doors and windows with *miskithu jaali*. The exterior and the interior are simple. The original mosque was probably built using coral stone, but the mosque has been modified with plasterwork to such an extent that it cannot be identified as a coral stone mosque.

Olhuvelifushi Old Mosque – *c*1900

The Old Mosque, Olhuvelifushi (Ghaviyani Atoll), has no clear date of construction, although there is a record of an ancient mosque built by Samiya Faashana (1620–1648). It is a small mosque constructed from coral rubble masonry and lime plaster along with timber. Numerous changes were undertaken around 2005.

Maradhoo Dhandivaru Mosque – *c*1900

Dhandivaru *Miskiy*, Maradhoo (Addu City), may be among the oldest mosques in the atoll and is associated with the historical oral story of Gaadhiru Kaleyfaanu. The story relates to Sanfaagey Koyya Beyya of Maradhoo who saw the ruins of the mosque and cemetery in the forest and rebuilt the mosque using *dhandi fan* – wattle and thatch. The date of construction of the present mosque is unclear and it has been renovated to such an extent that it is a modern structure in traditional mosque form. The mosque has a simple prayer hall with an open veranda at the entrance and a modern minaret. There are no decorations or carvings and the mosque is built using masonry and cement plastering.

Male' Bihurozukamana Mosque – 1919

Bihurozukamana *Miskiy,* Male', contains the mausoleum of Sultan Ghazi Mohamed Thakurufaanu. It existed when the Sultan died in 1585, was remodelled during the reign of Sultan Ghazi Hassan Izzuddin (1759–1767), and then extensively remodelled in 1919. It is a small mosque constructed from coral rubble masonry, lime plaster and timber.

Fuvah Mulah Old Friday Mosque – *c*1933

The Old Friday Mosque, Fuvah Mulah (Gnaviyani Atoll), was first erected by Sultan Mohamed of Dhevvadhoo (1672–1701) and rebuilt by Sultan Shamsuddin (1903–1933). It is constructed from timber, coral rubble masonry and lime plaster.

Dhidhdhoo Eid Mosque – 1940

The Eid Mosque, Dhidhdhoo (Haa Alif Atoll), was completed on 18 September 1940. Sultan Ghazi Mohamed Thakurufaanu (1573–1585) probably ordered the first mosque to be built on the site. It is constructed from coral rubble masonry and lime plaster.

Makunudhoo Small Mosque – *c*1940

The Small Mosque, Makunudhoo (Haa Dhaal Atoll), was built in 1940. It is a fine mosque constructed from timber, coral rubble masonry and lime plaster.

Goidhoo Old Mosque – *c*1940

The Old Mosque, Goidhoo (Baa Atoll), was probably completed around 1940. It is a fine mosque constructed from timber, coral rubble masonry and lime plaster. Now, however, it is in a dilapidated condition.

Coral Stone Mosques

Their temples are called Mesquites, which are well built of fair worked stone ... stone [is] raised from the shallows of the sea, where they get as much as they want, both length and thickness. It is polished, and of good grain, very white, but a little hard in some cases for cutting and working ... The manner of drawing up the stone from the sea is remarkable.

Francois Pyrard, French navigator, 1619

The *hirigaa* or coral stone mosque of the Maldives is the subject of this book. No doubt all traditional mosques have historic significance to the Maldives, but the coral stone mosque stands out because of its unusual construction technique and its technical and artistic finesse. It is the type that best illustrates the skills of the Maldivian people. In these mosques, coral stone is used in a unique dry-assembly construction technique similar to traditional joinery. This technique, coupled with artistic ornament using coral carvings, lacquerwork, timber carvings and calligraphy, is not seen in any other part of the world.

After comparison and review of the properties with similar attributes from the World Heritage list, the Tentative List and other databases using the UNESCO/WHC recommended frameworks, it is apparent that the coral stone mosques of the Maldives are also universally significant.

- Universally significant in terms of the fusion of cultures, because there are no structures in the world that illustrate and represent all the main regions of the Indian Ocean in one building and certainly not to the extent displayed in these mosques.
- Universally significant in the way coral stone is used for construction. All the coral stone structures in the Indian Ocean region use coral stone masonry set in mortar, whereas the Maldivian mosques are the only structures where coral stone is shaped and assembled using a dry technique. They also stand apart in the quality of workmanship and decoration by which parallels cannot be drawn with any other in the world.

- Universally significant because there are no identical structures in any other regions during any historic period. These structures were made during a specific period of time and the techniques and crafts have disappeared.
- Universally significant because there are no mosques in the world identical in their spatial form, type of construction and decoration to the coral stone mosques of Maldives.

There are 20 known coral stone mosques built between 1403 and 1815 distributed throughout the country. The size of the existing coral stone mosques varies from 10 to 270 square metres. Even though these structures span a period of more than 400 years, there are few differences in plan form or style and no regional differences. Specialist master mosque builders or master carpenters commissioned by the sultans must have travelled throughout the country.

The typical coral stone mosque is built on a raised coral stone plinth made of shaped coral stone blocks with decorative mouldings and entered by ascending stairs. The main structure is an interlocking system integrating walls, columns, doors, ceilings and the roof structure. The construction technique, resembling the system of interlocking dry joinery, is called 'coral stone carpentry' and it enables the whole mosque to be reassembled.

Similar to traditional mosques the typical spatial plan of the coral stone mosque is symmetrical, laid out on an axis with a rectangular prayer hall, occasionally with columns, and a combination of the veranda-like antechambers called *fenda* or *dhaala* attached to it on the eastern, northern and southern sides. The qiblah wall on the west side does not have a mihrab niche but often has mihrab chambers containing the mimbar: a raised pulpit from where the Imam delivers speeches. The main entrance with rising stairs is at the eastern side occasionally with secondary entrances on the sides. On both sides of the steps are timber ledges, which are used as seating for the elderly and the disabled.

The plan has four variations:

1. Mosques without a mihrab chamber and without side *fenda/dhaala* (eg, Hulhumeedhoo Fandiyaru mosque, 1586)
2. Mosques without a mihrab chamber but with side *fenda/dhaala* (eg, Ihavandhoo Old Friday Mosque, 1701)
3. Mosques with a mihrab chamber but without side *fenda/dhaala* (eg, Fenfushi Old Friday Mosque, 1692–1701)
4. Mosques with a mihrab chamber and with side *fenda/dhaala* (eg, Isdhoo Old Mosque, 1701).

Similar to many traditional mosques, the roof form is tiered, hipped and steeply sloping with large overhangs. Some mosques have two tiers while others have three. The roof was originally finished with plaited coconut thatching or Indian clay roofing tiles. The mosque has a coffered timber ceiling with a central stepped recessed area called a *laage'*. The ceiling is raised from the main beams using turned timber supports called *thona*. The door system, which has a uniquely shaped arched doorway and a sliding mechanism, is integrated into the main structure, taking secondary loads. The windows have open diagonal lattice timberwork called *miskithu jaali* (mosque lattice). Variations of the latticework are found between external and internal windows and differ in each mosque.

The external decoration of the coral stonework includes mouldings and surface carvings of varying levels of sophistication based on geometric and organic patterns. The top part of the stone plinth has mouldings and carvings with bands of arabesque floral and interlaced patterns. The middle has coffered geometric panels with a symbolic door in the centre of the panels, and the bottom has a series of simple mouldings. The walls and sides of the stairs are decorated with bands of interlaced foliage knotwork. Internal ornament includes timber carvings and traditional lacquerwork on flat surfaces called *laafenkurun* and turned timber called *liyelaajehun*. Both carvings and lacquerwork use arabesque floral patterns, geometric patterns and interlaced foliage knot patterns. A high level of attention is paid to the decoration of the *laage'*, roof beams, the mihrab and the entrance doors. The symbolic parabolic arched sliding door with a key and columns on both sides is found throughout the timber and coral decoration. The iconography of the ornament is based on floral, organic and geometric patterns borrowed from both the pre-Islamic and Islamic periods of the Indian Ocean region. Calligraphic decorations are usually quotes from the Quran and Hadith of the Prophet Mohamed (PBUH) in Thuluth Arabic script and also serve to identify the building and the patrons.

The 20 known coral stone mosques are found throughout the country from the northernmost Haa Alif Atoll to the southernmost Addu City. They are in differing states in terms of authenticity and original condition. Out of the 20 mosques, eight are presented to relate the story of the coral stone mosque. They have been selected taking into account the artistic and architectural value, their representation of the typologies and their representation in terms of size and time period. The selected mosques are as follows:

1. Hulhumeedhoo Fandiyaru Mosque, *c*1586
2. Male' Old Friday Mosque, 1658
3. Ihavandhoo Old Friday Mosque, 1701
4. Fenfushi Old Friday Mosque, 1701
5. Isdhoo Old Mosque, 1701
6. Meedhoo Old Friday Mosque, *c*1705
7. Male' Kalhuvakaru Mosque, 1789
8. Male' Eid Mosque, 1815.

Hulhumeedhoo Fandiyaru Mosque, *c*1586

Hulhumeedhoo Fandiyaru Mosque, in the Koagannu area on the island of Hulhumeedhoo, Addu City, was probably built around 1586 during the reign of Sultan Ibrahim III, also known as Kalaafaanu (1585–1609) and is still in use as a mosque today. It is the earliest and smallest surviving coral stone mosque that is in good authentic condition. Even though the mosque is basic in its artistic sophistication with no coral carvings, it is significant to show the evolution of the mosque design and its construction techniques.

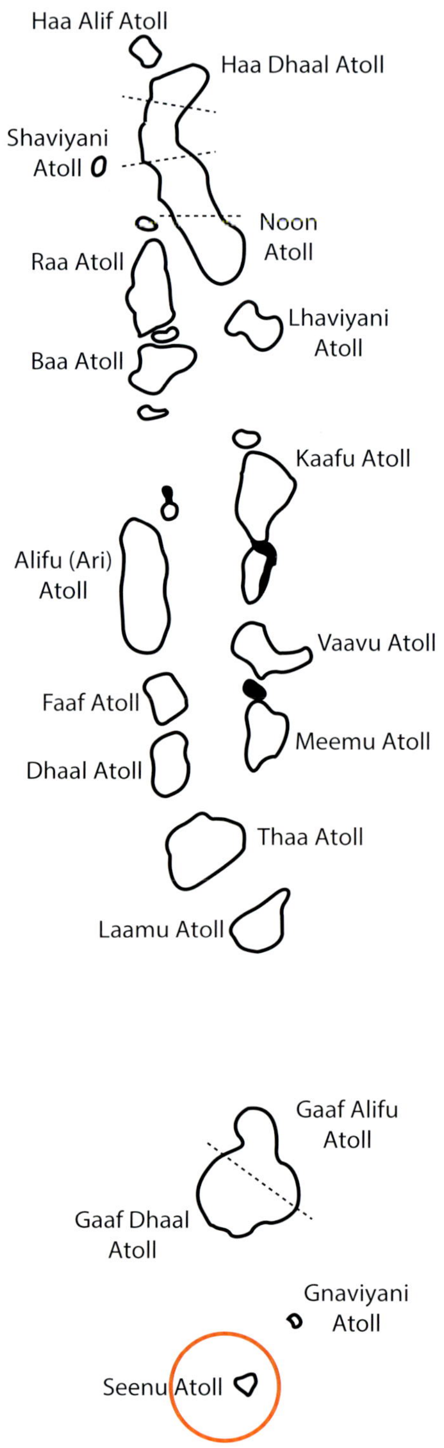

opposite

Location of Hulhumeedhoo

right

Location of Hulhumeedhoo

Fandiyaru Mosque

Location

Address: Koagannu, Hulhumeedhoo, Addu City, Maldives
Geographic coordinates: 0° 34' 51.6" S, 73° 13' 42" E

Hulhumeedhoo is in Addu City and about 528 kilometres south of the capital island Male'. It has a population of approximately 6,050. Traditionally, Hulhumeedhoo was famous as a centre for religious learning and for its many great religious scholars including Dhombeyya (Al Fageeh Al Sheikh Al Hafiz Ibrahim Thakurufaanu) who knew the Quran by heart, specialised in Unaani medicine and was also a master craftsman. It was also renowned in the region for its great traders and travellers.

The Koagannu area has the largest and the oldest cemetery in the Maldives with more than 500 coral stone tombs, a sheltered mausoleum and 15 open mausoleums. It had six small mosques, but only four remain. They are Koagannu Mosque *c*1397, Boadha Mosque *c*1403, Athara Mosque *c*1417 and Fandiyaru Mosque *c*1586. The largest coral stone tombstone is on this site and is 2.4 metres tall.

opposite
The Koagannu area of Hulhumeedhoo

right
Tombstones made from *hirigaa* coral stone

far right
The largest tombstone in the Maldives, found on Hulhmeedhoo, is 2.4 metres high and made from one piece of coral stone

History

Fandiyaru Mosque was probably built around 1586 during the reign of Sultan Ibrahim III. Koagannu was traditionally a remote part of the island: sometimes small mosques are located in remote areas to be used for I'tikaf, a days-long retreat to a mosque for prayers.

It is not clear who built the mosque but traditionally mosques in Koagannu were built and maintained for generations by different families from Hulhumeedhoo. Fandiyaru Mosque was traditionally maintained and looked after by the family of Sultan Ghazi Mohamed Bodu Fandiyaaru Thakurufaanu. Now it is maintained by the state and is still used as a mosque.

Koagannu area originally did not have a boundary wall and the first was built during President Mohamed Amin Didi's era, *c*1953. Even though the dates are indeterminate, it is evident that the mosque has gone through various stages of restoration and renovation. The roof is the most visible example.

Description

Mosque capacity: 10 people.
Approximate mosque area: 12 square metres.

The Fandiyaru Mosque is a simple building with a small minaret and an open mausoleum next to it. Stepping-stones from the mosque lead to a well used for ablutions. The building has a small prayer hall and open verandas on three sides. Three sides of the prayer hall have doors and the mihrab wall has no openings or ornament.

top

Present interior of the prayer hall

bottom

Present exterior

The mosque is a structure built on a low coral stone plinth with coral walls, timber doors, timber ceiling and a simple roof. The roof is two-tiered and has a simple coffered ceiling. Even though the mosque has roofing sheets now, originally it would have been coconut thatching. The arched doorway has sliding timber doors. The coral stone plinth has simple mouldings and does not have ornamentation in terms of external carvings or interior decoration. The coral stone plinth, walls, ceiling, doors and roof structure form an integrated structure brought together using an interlocking dry joinery system. Originally the floor finish would have been traditional reed matting and by night oil lamps would have lit the mosque.

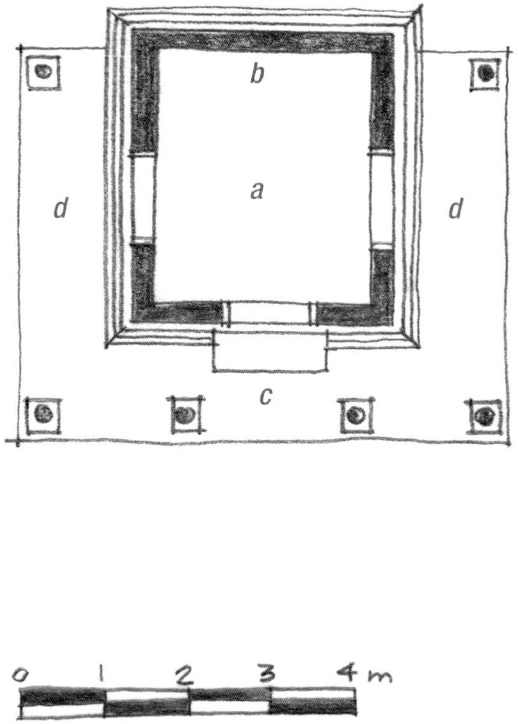
b
a
d
d
c
0 1 2 3 4 m

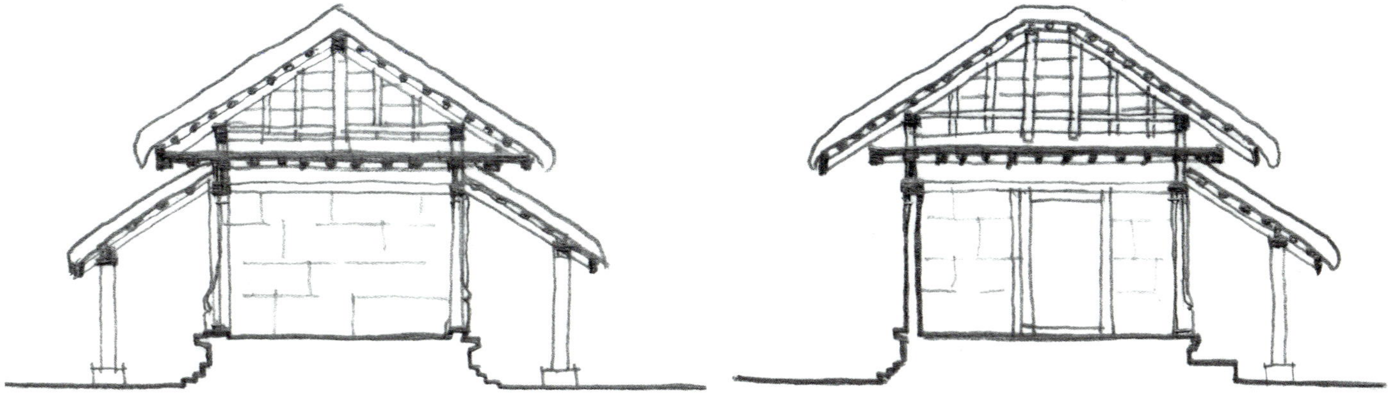

The authors' views of the original mosque

opposite top

Elevations

opposite bottom

Floor plan:

a prayer hall

b mihrab wall

c entrance

d *fenda*

top

Sections

right

Detail of timber wall plate

far right

Detail of timber sliding door

overleaf

The compound

opposite

Front without the roof structure

right

Interior of the prayer hall

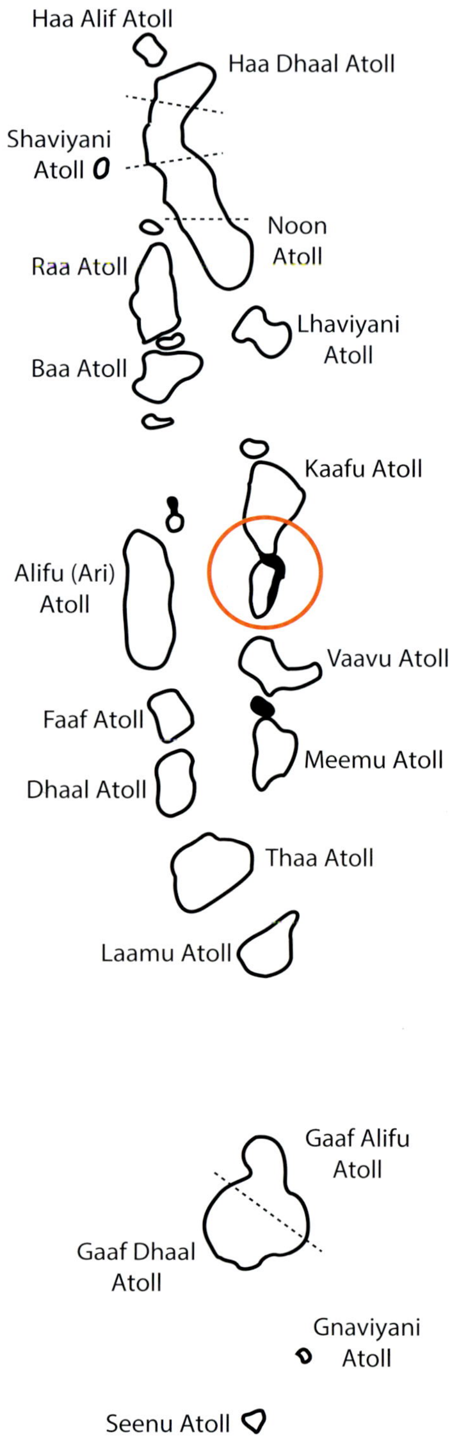

Male' Old Friday Mosque, 1658

Male' Old Friday Mosque and its compound comprise the most important heritage site in the country. It is also the biggest and one of the finest coral stone buildings in the world. It represents the best in terms of artistic excellence and the coral stone construction technique termed 'coral stone carpentry'; a sophisticated interlocking assembly technique without mortar. Male' Old Friday Mosque and its compound has been included in the UNESCO World Heritage Tentative List since 2008.

The present mosque was built in 1658 on the old part of the capital island of Male', during the reign of Sultan Ibrahim Iskandhar I. It replaced the original mosque built in 1153 by the first Muslim Sultan of Maldives and continues to serve not only as one of the most important places of worship, but also as a reminder of this country's great ancestral heritage. The mosque is in its original condition except for the corrugated metal sheet roof covering and the floor covering. The mosque compound is in reasonably good condition apart from the missing original gateway and part of the old cemetery. New landscaping alien to traditional mosque compounds compromises the original ambience.

opposite

Location of Male'

right

Location of Male' Old Friday Mosque

Location

Address: Fareedhee Magu, Henveiru, Male', Kaafu Atoll, Maldives
Geographic coordinates: 4° 10' 40.77" N, 73° 30' 44.57" E

Male' is the capital of Maldives and covers an area of about 200 hectares. With a population of over 100,000 it is among the most dense and smallest capital cities in the world. Male' has been the centre for politics, commerce and administration dating as far back as the Buddhist period, but sadly very little of its past heritage remains visible. Male' Old Friday Mosque is one of the few surviving structures that provides physical evidence of its ancient past. It is located on Fareedhee Magu in the older part of Male' near the main inner harbour, the old palace, the old bazaar and other important administrative buildings.

History

1153 Construction of the first mosque by the first Muslim Sultan, Sultan Mohamed Bin Abdullah. Details of this mosque are not known.

1338 Renovation of the mosque by Sultan Ahmed Shihabuddin.

1658 Construction of the present mosque by Sultan Ibrahim Iskandhar I. The master carpenters were Ali Maavadikaleyge' and Mahmud Maavadikaleyge' from Kondey, Gaafu Alifu Atoll and the principal calligrapher was Chief Justice Al Faqih Al Qazi Jamaludeen. The building took about two years to construct.

1676 Construction of the minaret and the main gateway by Sultan Ibrahim Iskandhar I after his return from Hajj in 1668. The gateway was the first place of formal education in the country.

1912 Restoration of the mosque and minaret by Sultan Mohamed Shamsuddin III. The roofing was changed from thatch to corrugated iron sheets.

1921 Construction of the boundary wall.

1940s Repair of the roof and addition of a new structure, a *haruge'*, to accommodate the increasing demand for space.

1964 Restoration of the roof and removal of the main entrance gateway. The original roofing was replaced with aluminium sheets and the roof structure was changed to teak wood. Loudspeakers were added to the minaret.

1980s Removal of the *haruge'* and part of the cemetery, adjustment of the mosque boundary and addition of landscaping.

1988 Scientific restoration work of the mosque with the assistance of UNESCO and the Indian Government. The first restoration project in Maldives.

Description

Mosque capacity: 300 persons.
Approximate mosque area: 240 square metres.

The mosque compound contains the mosque, a large minaret, three coral stone wells, a sundial and a cemetery surrounded by a boundary wall. The unique large drum-like minaret is made from plastered coral stone and tied with metal bands. It is painted in white with decorative calligraphy. The cemetery has coral stone mausoleums and tombstones of past sultans, princes and other dignitaries; the mausoleum of Sultan Ibrahim Iskandhar I is located here. The coral stone work within the cemetery comprises the finest collection in the country. The high-quality coral stone wells have stepping stones leading to the entrances and a sundial. Unfortunately, the compound no longer has well-established traditional trees.

The mosque compound with mausoleums and tombstones

opposite

A mausoleum

right

Detail of a mausoleum

far right

Tombstones

opposite

Existing front of the mosque

right

Side entrance

The mosque has a hypostyle layout with two prayer halls, six *dhaala/fenda* (antechambers), and two chambers for the sultan and his guards. There is a large mihrab chamber and the mimbar is in a corner of the mihrab chamber. The mosque can be entered through three entrances, each with rising steps. It is built on a highly decorated coral stone plinth using an interlocking dry joinery system to join together the coral stone wall, columns, beams, doors and roof structure. The roof is three-tiered with a modern metal roofing finish and has a highly decorated coffered ceiling, a central recessed *laage'* and 10 smaller recessed *laage',* all raised

opposite

Existing exterior of the mosque and water well

top

Newel post detail

bottom

Stairs

from the main beams by turned lacquered *thona*. There are 24 columns in the mosque, out of which 12 are made from coral stone. The qiblah board fixed to the mihrab wall is an exquisite piece of design and calligraphy. It displays quotations from the Quran and information about the sultan and how the mosque was built. With its fine exterior coral carvings and interiors with carved wooden doors, lacquerwork and calligraphy, the quality of workmanship is among the best in Maldives.

opposite

Dhaala

right

The central prayer hall

far right

Timber column head

opposite

Mihrab chamber and mimbar

right

A coral stone column

opposite and right

Laage'

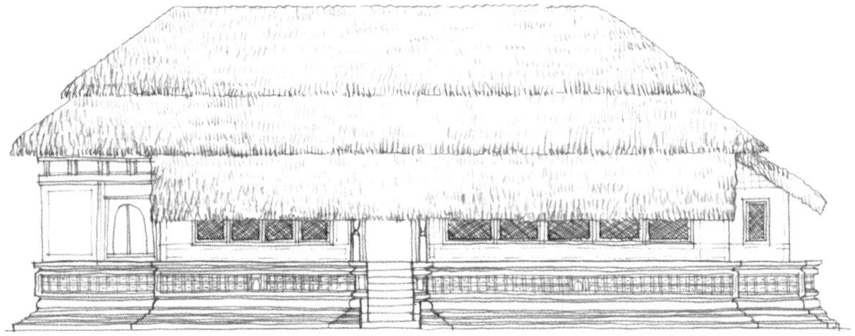

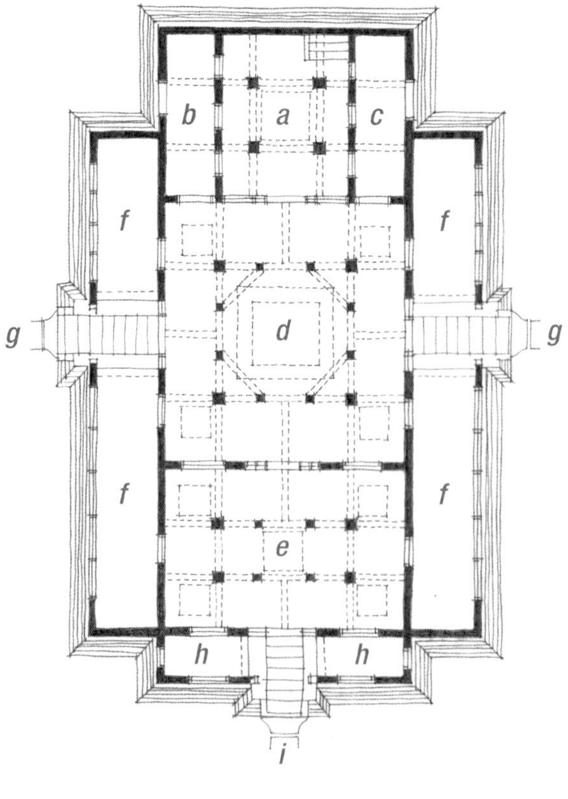
b
a
c
f
f
g
d
g
f
f
e
h
h
i

opposite top left

Side elevation

opposite top right

Front elevation

opposite bottom

Floor plan:

a Mihrab chamber

b Sultan's chambers

c Sultan's aides

d Central prayer hall

e Outer prayer hall

f Side *dhaala*

g Side entrances

h *Dhaala*

i Main entrance stairs

top right

Cross section

top far right

Long section

far right

Interior of the prayer hall

overleaf

The compound

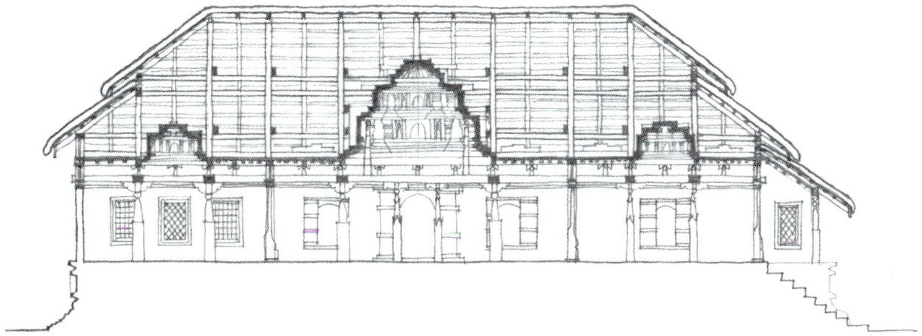

The authors' views of the original mosque

To appreciate the mosque in its original state one has to imagine the exterior with coconut thatch roofing, its main entrance gate, the minaret without metal railing, appropriate landscaping and without the modern buildings around it. The interior timber would have been lighter and brighter. Originally the floor would have been traditional reed matting and oil lamps would have lit the mosque at night. The following illustrations depict the mosque in its original state.

opposite top

Coral carvings on the side of the entrance door – seal of the Sultan Ibrahim Iskandhar

opposite bottom

Entrance detail

top

Coral carvings – Celtic style interlaced organic knot patterns – on the sides of the entrance stairs

right

Coral carvings – geometric coffered patterns – on the wall of the *dhaala*

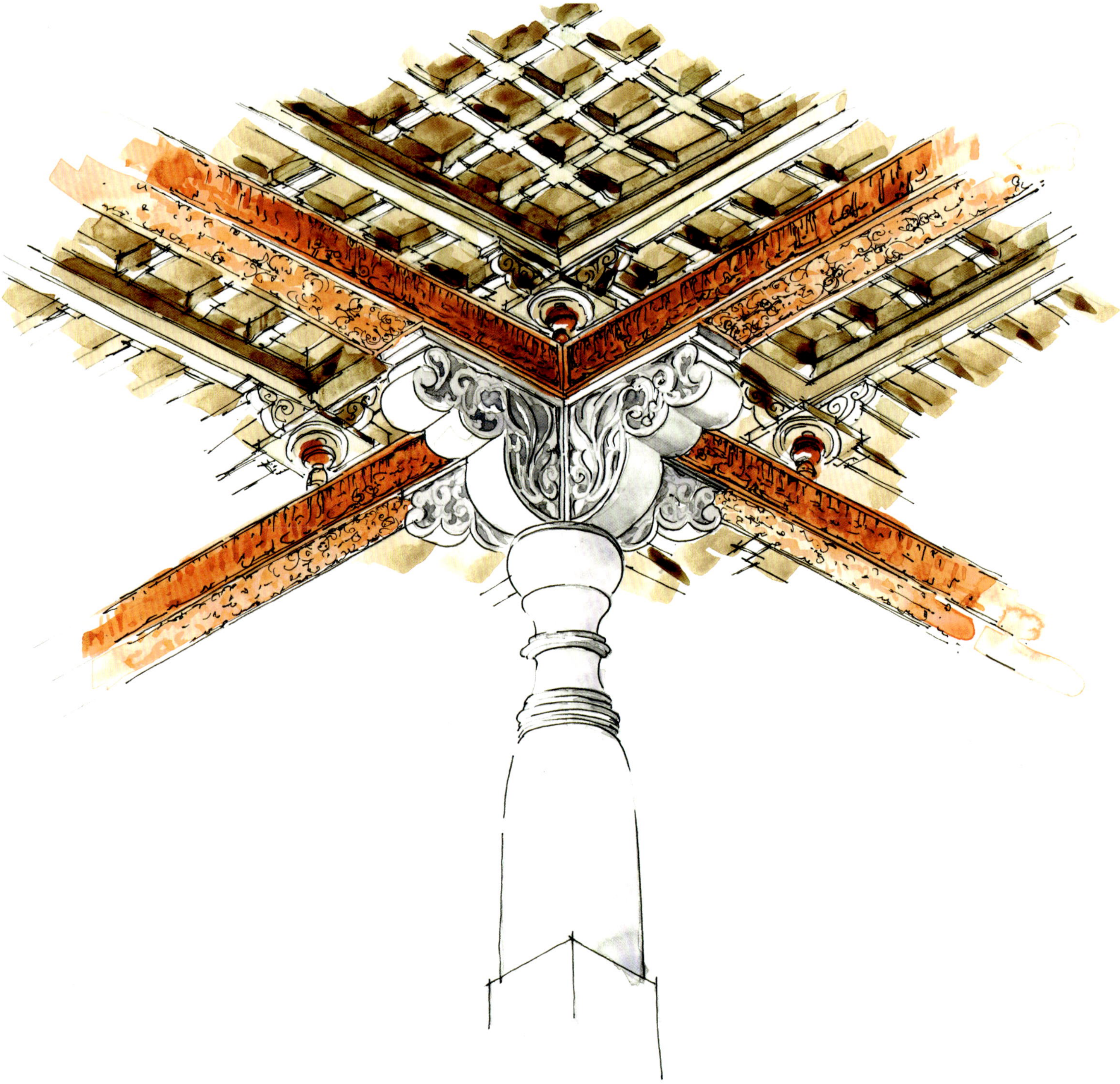

opposite

Detail of carved coral column head and lacquered timber main beams

right

Lacquered calligraphy panel on wall panel of the prayer hall, Jeli Thuluth style

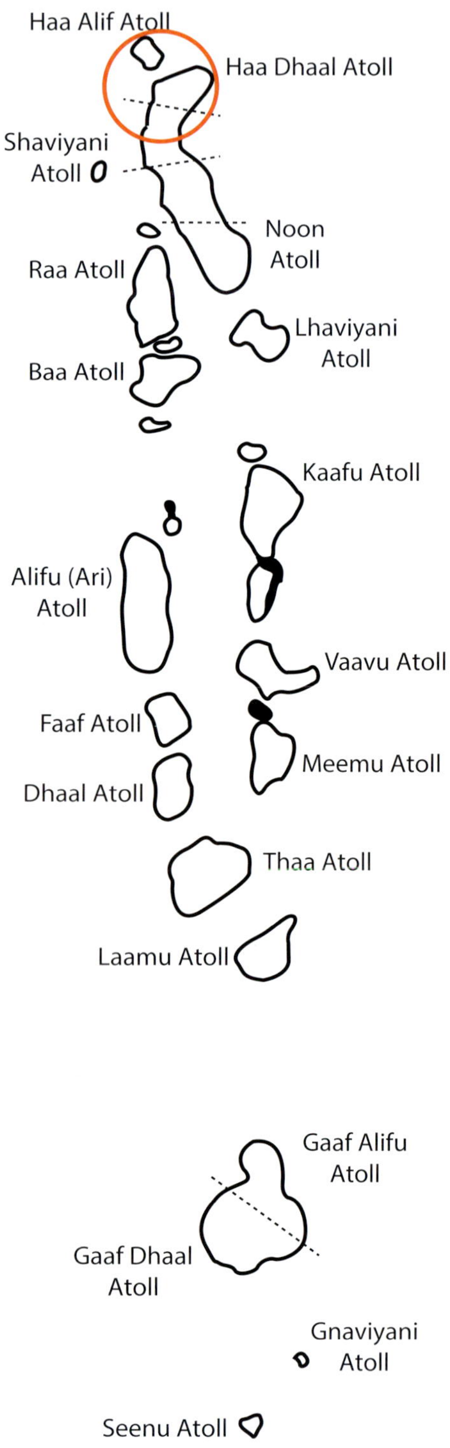

Ihavandhoo Old Friday Mosque, 1701

Ihavandhoo Old Friday Mosque is the finest coral stone mosque in the north of the Maldives in terms of its decoration and workmanship. It is a small mosque with *dhaalas/fendas* on three sides and has high-quality coral stonework and intricate interior details. The mosque compound has coral stone tombstones with high-quality carvings displaying rare patterns. The old mosque survives in good condition but unfortunately major additions make it difficult to appreciate its original beauty.

Standing at the end of the main road of Ihavandhoo Island in Haa Alif Atoll, the mosque was completed on 16 December 1701, during the reign of Sultan Ibrahim Muzhiruddin (1701–1705). It is still used today to serve its community as a centre for prayer.

opposite

Location of Ihavandhoo

right

Location of Ihavandhoo

Old Friday Mosque

Location

Address: *Miskiy* Magu, Ihavandhoo, Haa Alif Atoll, Maldives
Geographic coordinates: 6° 57' 17.33" N, 72° 55' 38.33" E

The island of Ihavandhoo is in the northernmost atoll of Haa Alif, 314 kilometres north of the capital island Male'. The island's area is around 61 hectares with a population of approximately 3,000. Because of its location, the old harbour had historical significance to early travellers as a gateway to Minicoy Island and the western coasts of India, Africa and Arabia. In ancient times, the island was famous for its coir rope but currently the economy depends on fisheries. The history of the island is linked to the Dhiyamigil dynasty and to the family of Sultan Ghazi Mohamed Thakurufaanu, the Maldivian hero who defeated the Portuguese occupying Maldives. The island's oral history relates a story of a religious man from 'Habush Karaa' (present-day Ethiopia) who settled on the island. The people of the island are strong and hard working with many of their cultures associated with the Kerala and Malabar coastal regions.

History

According to the oral history there were older mosques on the island preceding Ihavandhoo Old Friday Mosque but it is not recorded whether there were any in the same location.

1701 The present mosque in its original form was completed on 16 December and bears the seal of Sultan Ibrahim Muzhiruddin.

1950s During the 1950s, the roofing was upgraded from coconut thatching to Indian clay roofing tiles and the *dhaala* was extended with timber lattice windows (*miskithu jaali*). The floor would have had a new screed. The short minaret was also part of the upgrade.

Existing exterior of the mosque

1990s The first electrification and linoleum floor finishes were introduced.

2005 The roof was changed from clay roofing tiles to metal profile sheets. The *dhaala* was extended without raising the floor level, and timber lattice windows and doors were added. The floor was changed to ceramic tiles. Modern facilities like electric lights, fans and sound systems were added.

Description

Mosque capacity: 40 persons.
Approximate mosque area: 38 square metres.

The mosque compound contains the mosque, a short minaret, an octagonal water well, a mausoleum and tombstones, all built from coral stone. The short minaret, about 1.5 metres high with six rising steps, was used in the past by the muezzin to call five times daily for prayers. The octagonal water well built using coral stone slabs still provides sweet water for ablutions and drinking. The cemetery has a number of old coral stone tombstones and a mausoleum. The tombstones with their fine carvings and Arabic dates are in good condition but the mausoleum no longer has a roof. Some of the tombstones belong to members of the Sultan Ghazi Mohamed Thakurufaanu family and the wazir of the Dhiyamigil dynasty. The original mosque compound probably did not have a boundary wall but a new boundary wall with three entrances now surrounds the mosque. Presently there are no trees or plants in the mosque complex, but until recently there were many frangipanis, jasmine and other flowering plants.

right

The mosque compound with mausoleums and short minaret

top and far right

Tombstones

right

Detail of the raised plinth

top

Existing stairs

far right

Newel post detail

opposite

Existing exterior of the old mosque

opposite

The central prayer hall

The mosque consists of the old original mosque built using coral stone carpentry techniques and a new extension erected with masonry construction. The old mosque is a typical small coral stone mosque with a prayer hall and *dhaala* (antechambers) on three sides. There is no mihrab chamber and the mimbar is near the corner of the mihrab wall. Typical of all mosques, it is entered opposite the mihrab side by stairs. The mosque is built on a raised coral stone plinth with coral stone walls, timber doors and windows.

The roof form is two-tiered and the timber roof structure and ceiling are constructed mainly in teak. There are no columns in this mosque so the roof structure is supported by the combined wall and door systems. The coffered ceiling has a central recessed *laage'* supported by a *thona*. With its fine coral carvings, entrance steps, carved wooden doors, lacquer interiors and calligraphy, the quality of the workmanship is among the best to be found outside Male'. The floor was finished with traditional reed mats and in former times oil lamps were used for lighting. The last new additions were built in 2005 to cope with the increasing demand for more space and better facilities. These were done using modern masonry, metal roofing, lattice windows, ceramic floor tiles and new electric fittings but the shape of the original roof was maintained.

opposite

Laage'

right

Laage' ceiling

far right

Calligraphy panel

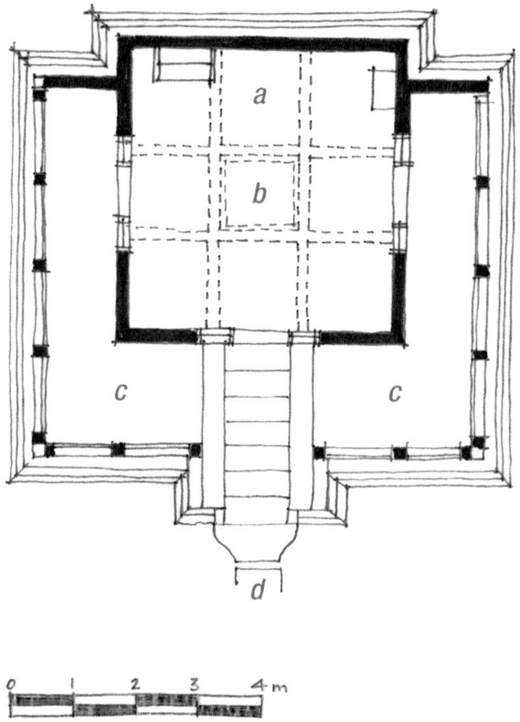
a
b
c
c
d
0 1 2 3 4m

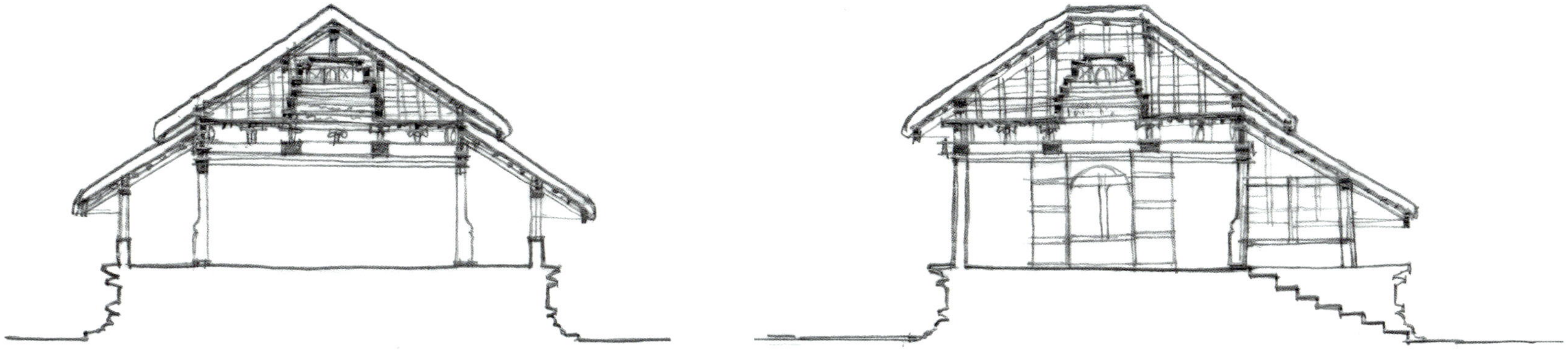

The authors' views of the original mosque

opposite top

Elevations

opposite bottom

Floor plan:

a Mihrab wall

b Prayer hall

c *Dhaala*

d Entrance

top

Sections

far right

Coral carvings – arched door design with keys – on the plinth

overleaf

The compound

The old mosque is practically inside a new mosque, so to appreciate the old mosque in its original glory one has to imagine the mosque without its extensions. The mosque has to be pictured with a coconut-thatched roof, the whitewashed minaret, coral stone well, tombstones and appropriate landscaping and without the modern boundary wall and buildings around it. The interior timber would have been lighter and brighter with traditional reed floor mats. The following artistic sketches depict the mosque in its original state.

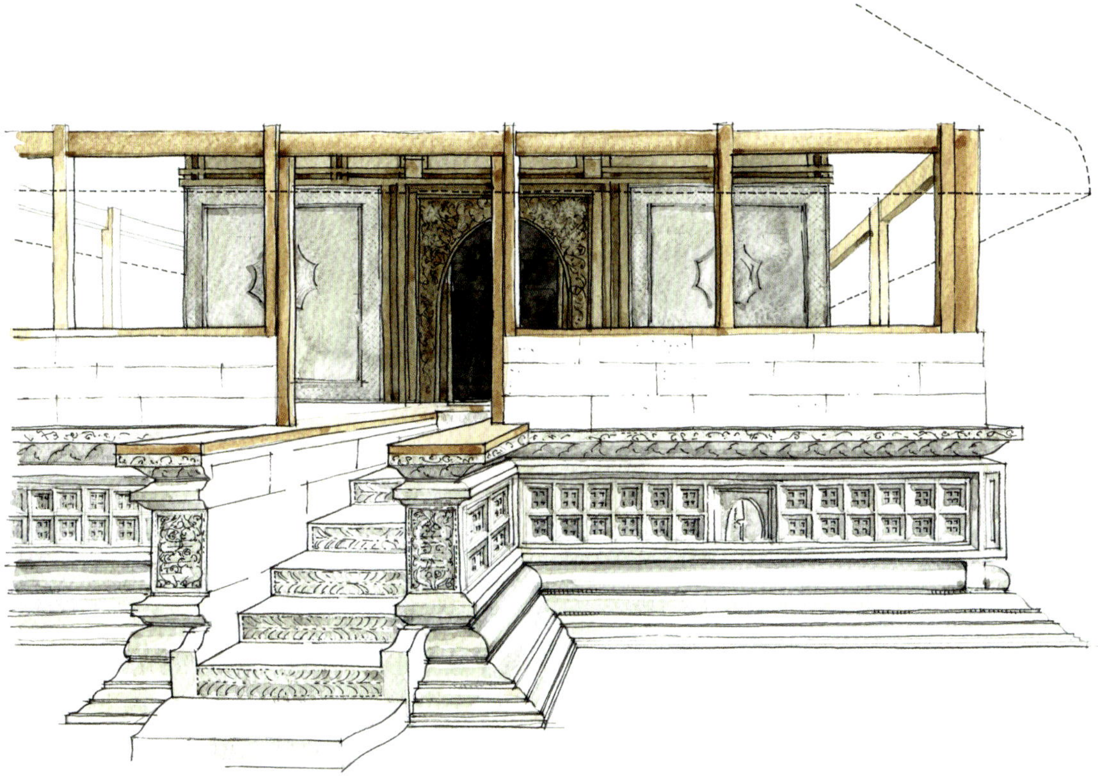

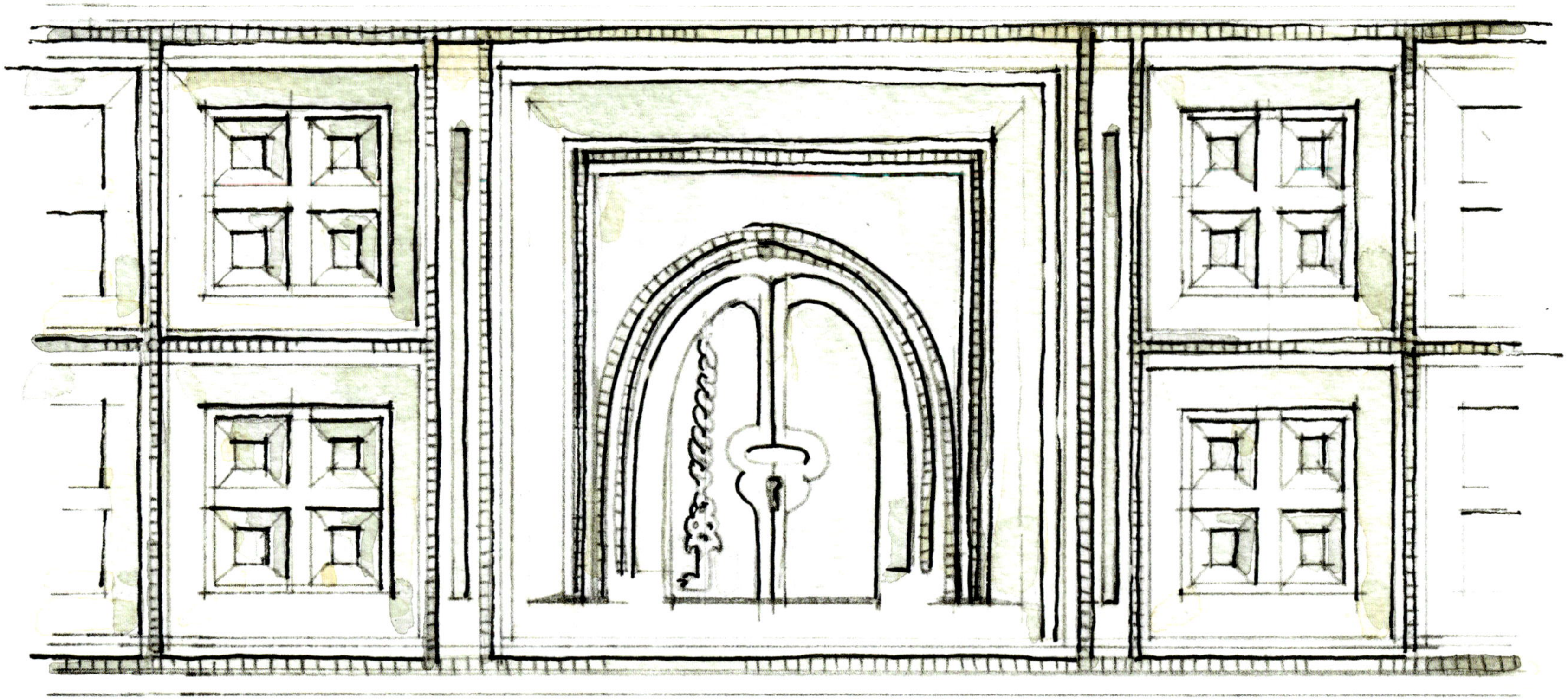

opposite top

Coral carvings – finite organic knot patterns – on the entrance stairs newel post

opposite bottom

Entrance detail

right

Coral carvings – South Asian style Arabesque patterns – on the plinth coping

far right

Coral carving – seal of the Sultan – on the sides of the entrance

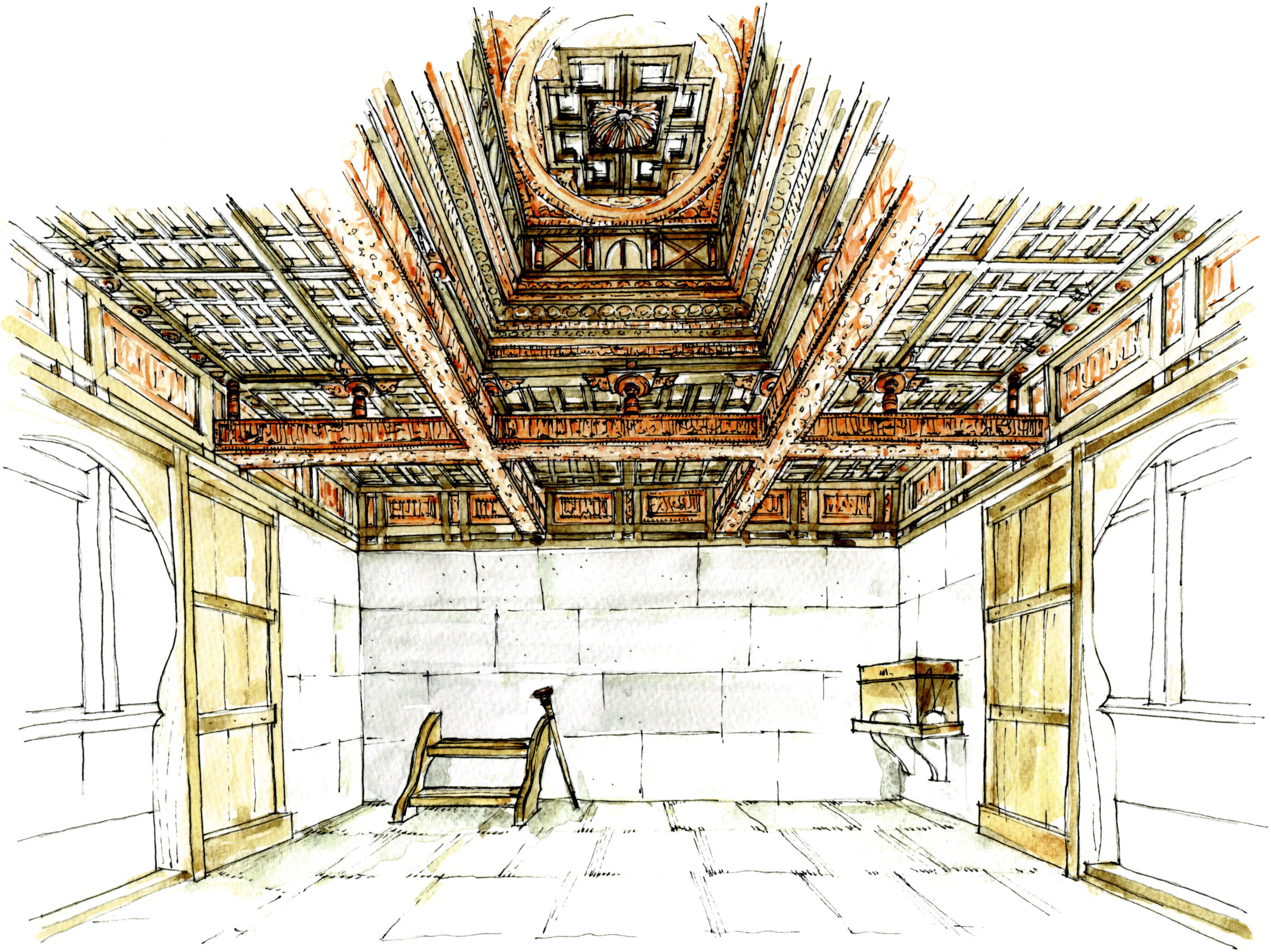

opposite
Interior of the prayer hall
right
Lacquered calligraphy panel on a wall panel of the prayer hall, Jeli Thuluth style
far right top
Thona – turned lacquered short post in the prayer hall
far right bottom
Lacquered design – finite floral palmette pattern – on the *laage'* ceiling

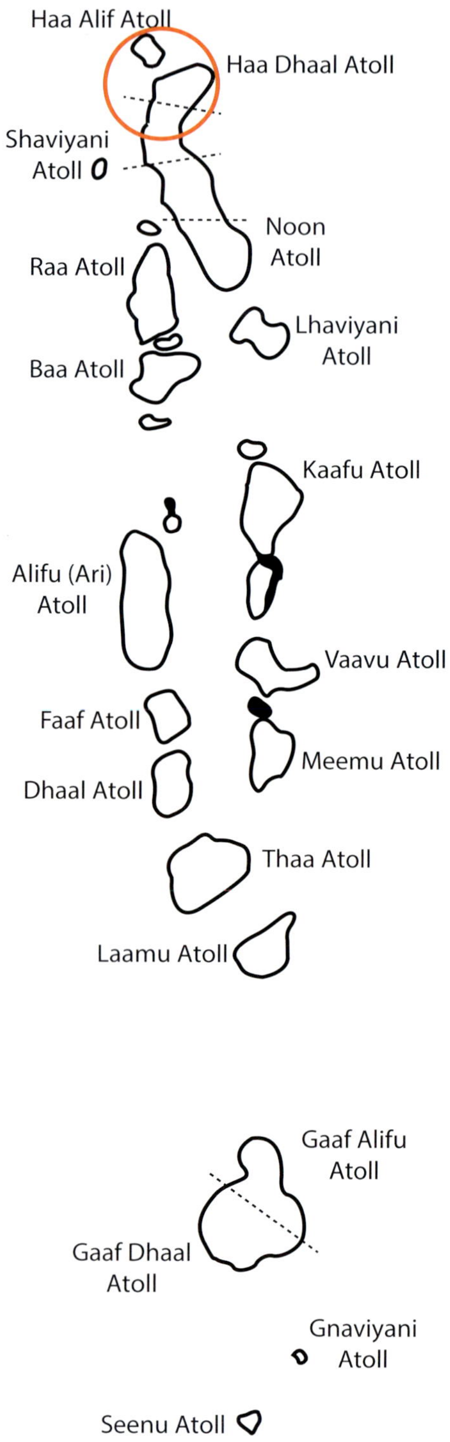

Fenfushi Old Friday Mosque, 1701

Fenfushi Old Friday Mosque is another fine example of coral stone architecture in the Maldives, the coral carvings and detailed lacquerwork exemplifying the high-quality decoration and workmanship. It is part of the best-preserved coral stone mosque compound with a unique coral stone bathing tank, coral stone wells, a sundial and a large cemetery with tombstones of a high standard.

Situated at the centre of the island of Fenfushi, it was rebuilt during the reign of Sultan Mohamed of Dhevvadhu (1692–1701) on the site of an earlier mosque erected by Kalhu Kamana and is still used as a mosque today. Except for the metal roofing on the mosque and over the well and bathing tank, it is in its original condition.

opposite

Location of Fenfushi

right

Location of Fenfushi Old Friday Mosque

Location

Address: Hiriga Goalhi, Fenfushi, Alif Dhaal Atoll, Maldives
Geographic coordinates: 3° 45' 15" N, 72° 58' 35" E

The island of Fenfushi is in Alif Dhaal Atoll and is about 109.5 kilometres south of the capital island of Male'. The population is approximately 825. The island economy was based on coral and sand mining until the introduction of tourist resorts to the region. Fenfushi was famous for its coral stone and sand miners. The last coral stone carpenters to make coral stone tombstones are old and no longer practise the art. The history of the island is linked to the sultans of the Utheemu, Dhiyamigil and later Huraa dynastics, who provided many services and privileges to the island. Mariyam Kamba, queen of three sultans, was the daughter of Fenfushi Kuraha Hassan Kaleyfaanu. The children of Sultan Ghazi Hassan Izzuddin (1759–1767) also visited the island to see family. The sites of two palaces called Boduganduvaru and Kudaganduvaru are claimed to be on the island.

opposite

The mosque compound with mosque, cemetery and wells

right

Existing exterior of mosque

History

The present mosque was constructed during the reign of Sultan Mohamed of Dhevvadhu (1692–1701) on the site of an earlier mosque built around the early 1600s by Kalhu Kamana of the Utheemu dynasty. The structure is well preserved and intact and there are no records of any changes. Nevertheless, the roofing must have changed many times. Scientific studies and subsequent preservation of the mosque were conducted in 2000 and 2011. New metal roofing was added over the mosque, water well and bathing tank in 2000.

Description

Mosque capacity: 60 persons.
Approximate mosque area: 58 square metres.

The mosque compound contains the mosque, a large and rare coral bathing tank, coral stone wells, a sundial, a cemetery with many tombstones and a boundary wall. The coral masonry boundary wall surrounding the mosque has two entrances. The building has a hypostyle layout with three entrances, one main rising staircase at the front of the mosque and a prayer hall with no *dhaala*, or veranda-like antechambers. A mihrab chamber is present and the mimbar is in the corner of the chamber. The mosque is built on a highly decorated coral stone plinth with coral stone walls, columns, beams, doors and roof, all joined using

opposite left

The mosque compound with sundial

opposite right and right

Mosque wells

top

Bathing tank

an interlocking dry joinery system. The roof is two-tiered with a modern metal roofing finish and has a highly decorated coffered ceiling with a central stepped recessed *laage'*. The coffered ceiling and *laage'* are raised from the main beam using turned timber *thona* and the main columns are also made from timber. The plinth is decorated with mouldings and surface carvings of geometric and organic patterns. The interior has decorative carvings of organic patterns on the doors and delicate lacquer and calligraphic work. The accomplished workmanship of the exterior and interior is among the finest in the country. Originally, the floor would have been finished with traditional reed mats and the mosque lit by oil lamps.

opposite top and bottom left

Detail of the plinth decorations

opposite bottom right

Detail of the raised plinth

right

Existing exterior of the mosque

right

Existing stairs

opposite

The central prayer hall

opposite left

The central prayer hall

opposite right and right

Laage'

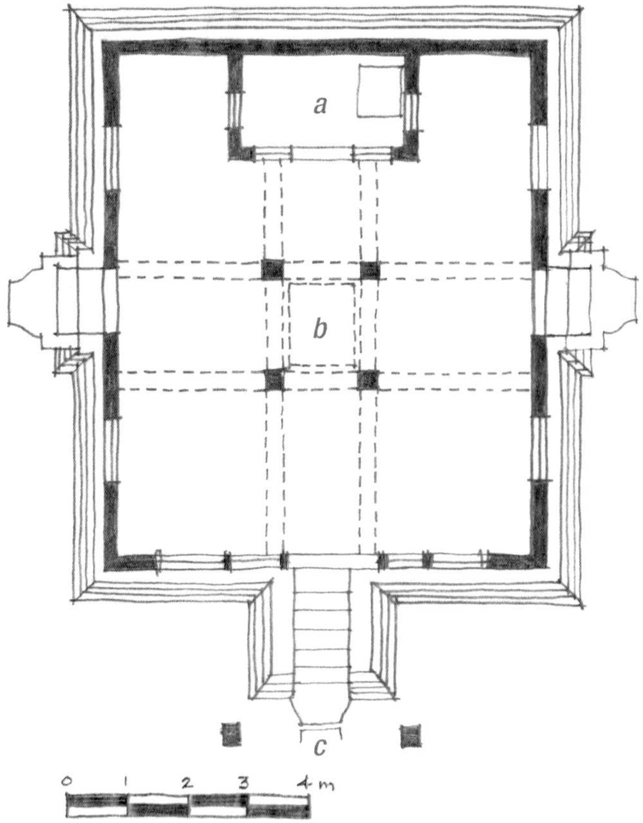
a
b
c
0 1 2 3 4 m

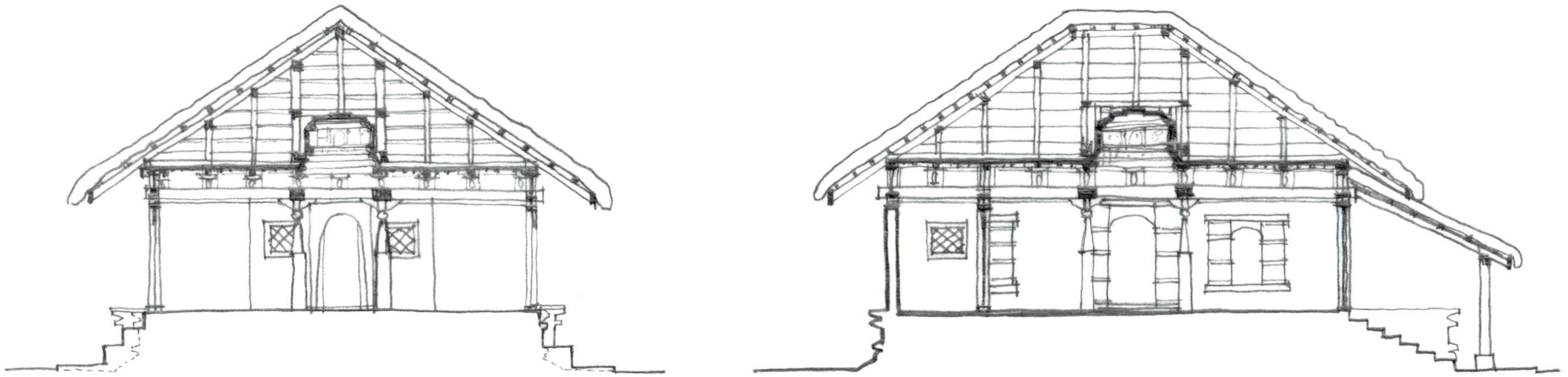

opposite top

Elevations

opposite bottom

Floor plan:

a Mihrab chamber

b Prayer hall

c Main entrance

top

Sections

far right

Coral carvings – symbolic parabolic door design – on the plinth

overleaf

The compound

The authors' views of the original mosque

Except for the new roofs, the mosque is in its original condition. To be appreciated in its original state, the exterior has to be imagined without a roof over the wells and bathing tank and with coconut-thatch roofing on the mosque itself. The interior timber would have been lighter and the floor would have had traditional reed matting. The following illustrations depict the mosque in its original state.

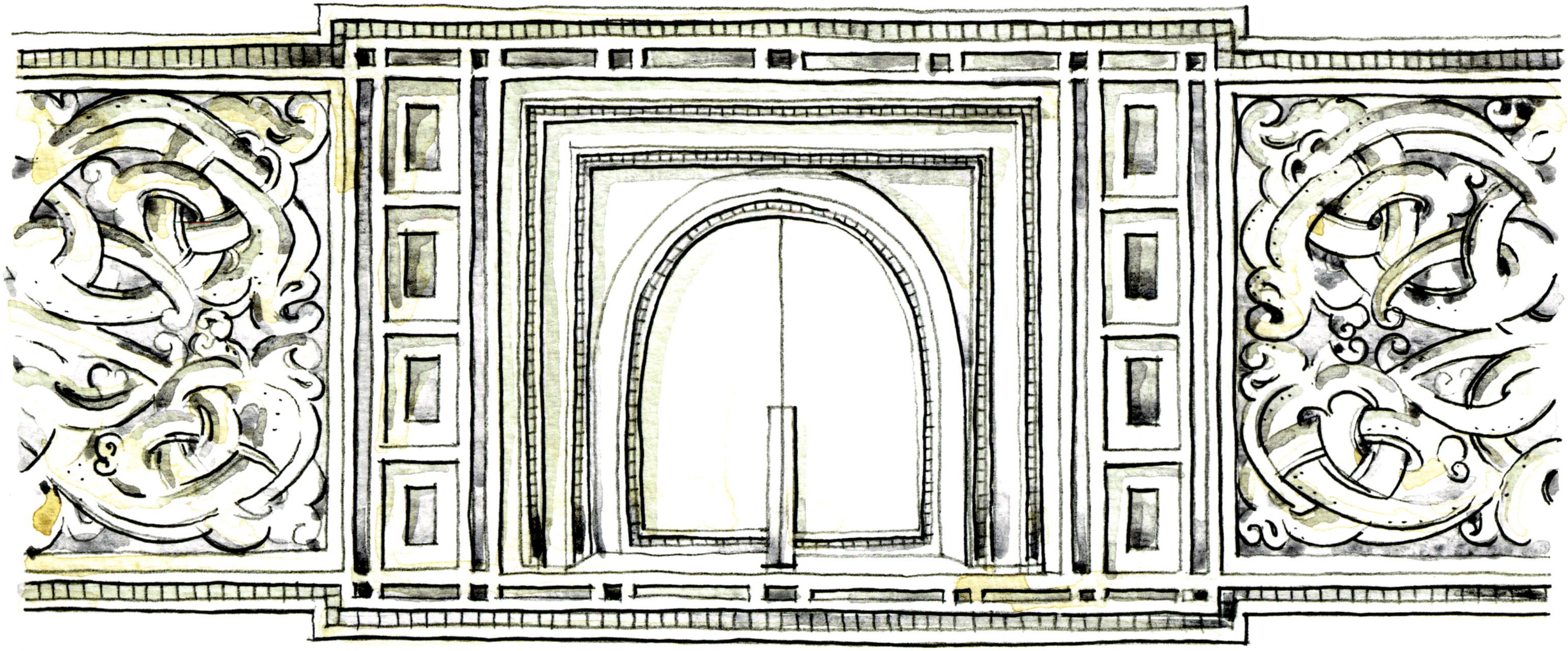

opposite

Entrance detail

right

Coral carving – finite floral pattern – on the plinth

opposite

Interior view of the prayer hall

right

Detail of lacquered timber ceiling of the central *laage'*

Isdhoo Old Mosque, 1701

Isdhoo Old Mosque is a beautiful small coral stone mosque with mihrab chamber, *dhaala* on three sides, and three entrances. This is the only mosque with an asymmetric floor plan and a special chamber. The mosque is historically significant because the special chamber was where the ancient royal copper chronicles dating from the 12th century – the *Isdhoo Loamaafaanu* – were kept. It has no coral carvings on the exterior, but high-quality interior ornament.

The mosque was built near the harbour area of Isdhoo in Laamu Atoll, in 1701 during the reign of Sultan Mohamed of Dhevvadhu. It is still used as a mosque today and has been restored and well-maintained with minimal changes. The mosque compound contains the mosque, old well and a cemetery with tombstones. There are no boundary walls.

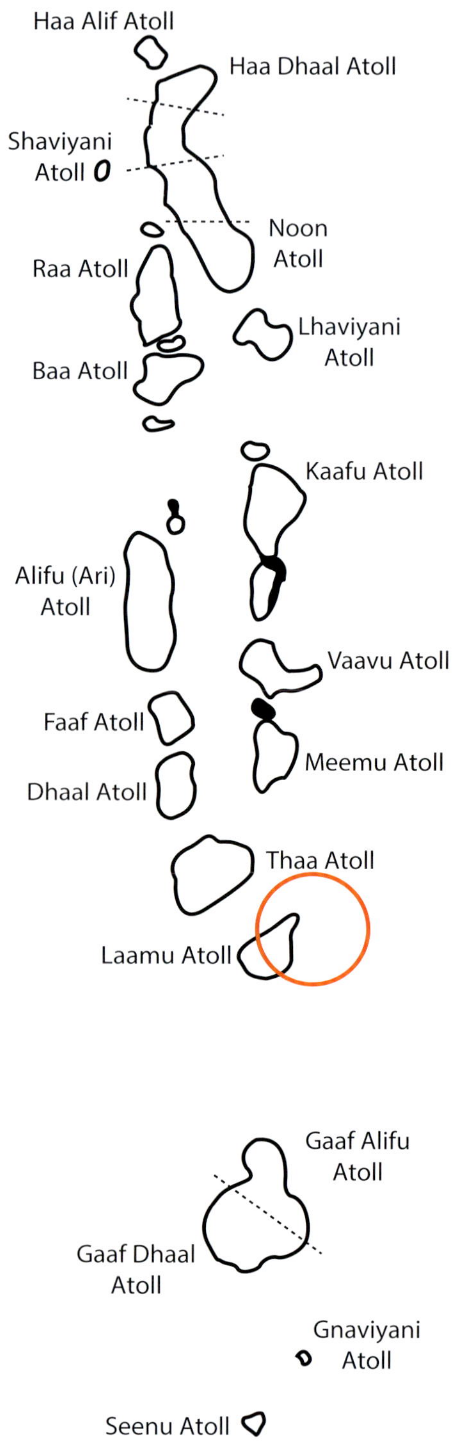

opposite

Location of Isdhoo

right

Location of Isdhoo Old Mosque

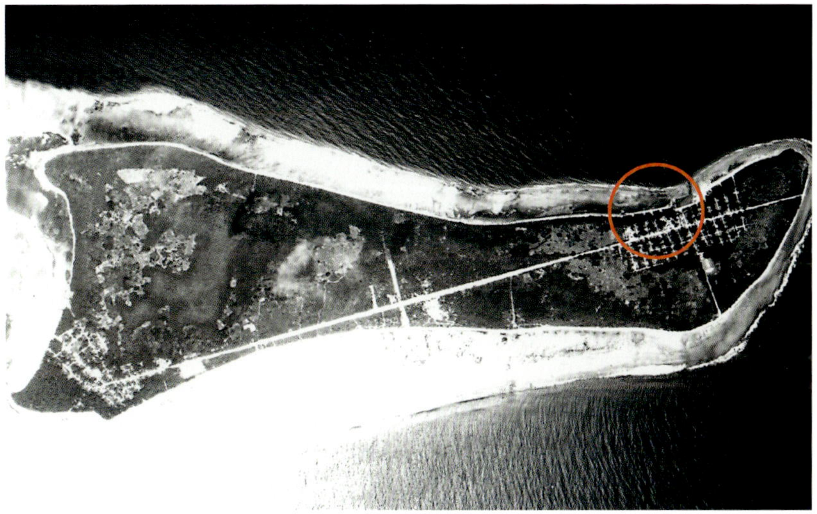

Location

Address: Isdhoo, Laamu Atoll, Maldives
Geographic coordinates: 2° 7' 10" N, 73° 34' 10" E

Isdhoo is part of a large island in Laamu Atoll. It is about 224 kilometres south of the capital island of Male' and has an area of about 160 hectares. The population is approximately 2,040. Isdhoo has important archaeological sites dating back to the Buddhist period, including Isdhoo Havitha, which is among the biggest in the country. The island also has connections to the sultans of the Isdhoo dynasty.

History

According to local oral history, a mosque was built here by a master carpenter named Ismail Maavadikaleyge' around 1657–58, using the leftover coral stone and timber from Male' Old Friday Mosque. The present mosque was built or remodelled by Sultan Mohamed of Dhevvadhu (1692–1701) and has survived well without many changes. Even though there are no records, the study of old photographs shows that the mosque has been restored numerous times with the roofing changing from coconut thatching to asbestos roofing sheet to modern metal sheets. Parts of the interior timberwork have been replaced and restored but evidence of a fire is still visible. The last record of restoration was in the early 2000s.

Description

Mosque capacity: 80 persons. Approximate mosque area: 75 square metres.

The mosque complex consists of the mosque, a water well and a small cemetery without a boundary wall. The tombstones in the cemetery are badly damaged and in poor condition.

opposite

The mosque compound with mosque, cemetery and wells

right

Tombstone

This mosque has a prayer hall with *dhaala* (veranda-like antechambers) on three sides, a mihrab chamber, a special chamber to keep the *Isdhoo Loamaafaanu*, and three entrances. It is the only mosque with an asymmetric floor plan. The mosque is built on a coral stone plinth with coral stone walls, four coral stone columns, timber doors and roof structure all brought together using an interlocking dry joinery system. The roof is two-tiered with a modern metal roofing finish and has a decorated coffered ceiling with a stepped recessed area called a *laage'* raised from the beam using *thona*. The timber doors have arched doorways with sliding doors and the windows use a timber lattice style called *miskithu jaali*. The coral stone platform is decorated with layers of mouldings but does not have fine carvings. The mosque has carved wooden doors, lacquerwork and calligraphy of good quality. Originally the floor finish would have been traditional reed mats and hanging brass or glass oil lamps would have lit the mosque.

opposite and right

Existing exterior of the mosque

far right top and bottom

Existing plinth and stairs

opposite

Detail of the stairs

right

Detail of the entrance door

far right

The central prayer hall

opposite

The central prayer hall

right

Ceiling with beam and *thona*

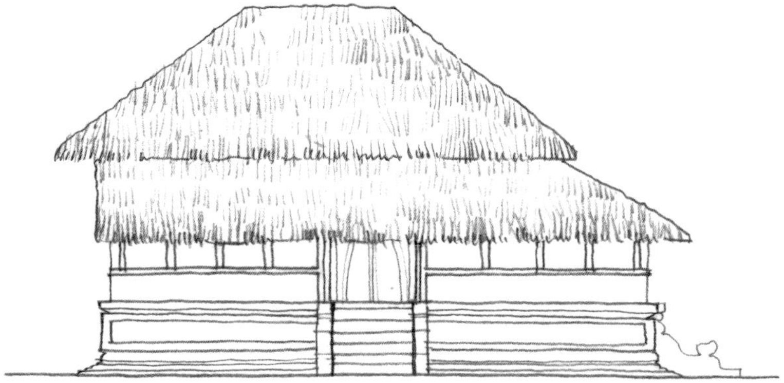

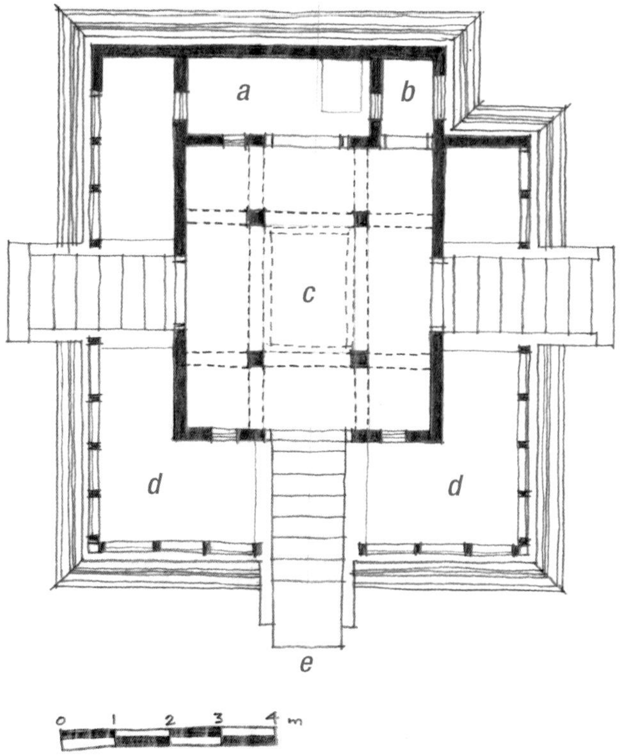
a
b
c
d
d
e
0 1 2 3 4 m

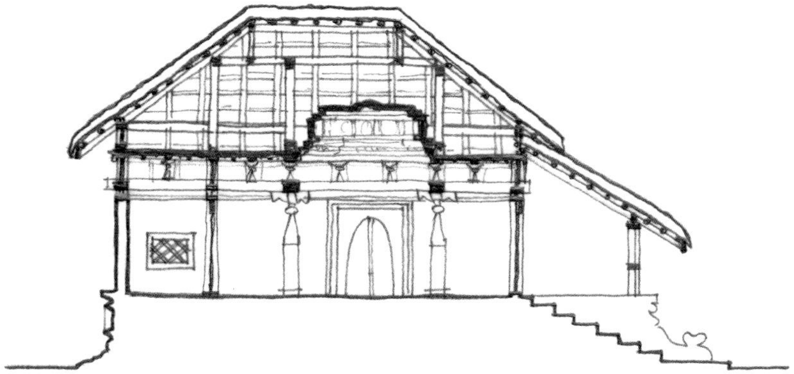

The authors' views of the original mosque

opposite top

Elevations

opposite bottom

Floor plan:

a Mihrab chamber

b Chamber for *Isdhoo lomaafaanu*

c Prayer hall

d *Dhaala*

e Main entrance

top

Sections

It is easy to appreciate the old mosque in its original glory. The mosque has to be imagined with coconut-thatch roofing, coral stone well and tombstones. The interior would have been the same but lighter and brighter with traditional reed floor mats. The following illustrations depict the mosque in its original state.

below

Entrance detail

far right

The compound

top

Detail of the the coffered ceiling, beam and calligraphy panel

middle

Lacquered calligraphy on the main beam, Jeli Thuluth style

bottom

Lacquered organic decorations of the main beam

opposite

Interior of the prayer hall

Meedhoo Old Friday Mosque, *c*1705

Meedhoo Old Friday Mosque is the only surviving coral stone mosque in the northern central region of the Maldives and the only one remaining with traditional Indian clay roofing tiles. It is a good example of a coral stone mosque with *dhaala* and a mihrab chamber, but the coral workmanship is simpler than in many other mosques. The cemetery has also survived well and has a collection of coral stone tombs with a high standard of carvings and craftsmanship.

The mosque is at the centre of the island of Meedhoo in Raa Atoll. The exact period of construction is unclear but according to local oral history it is around 300 years old and was probably built during the reign of Sultan Muzaffar Mohamed Imaduddin around 1705. It is still in use as a mosque today and is well maintained having experienced few changes since the early 1900s.

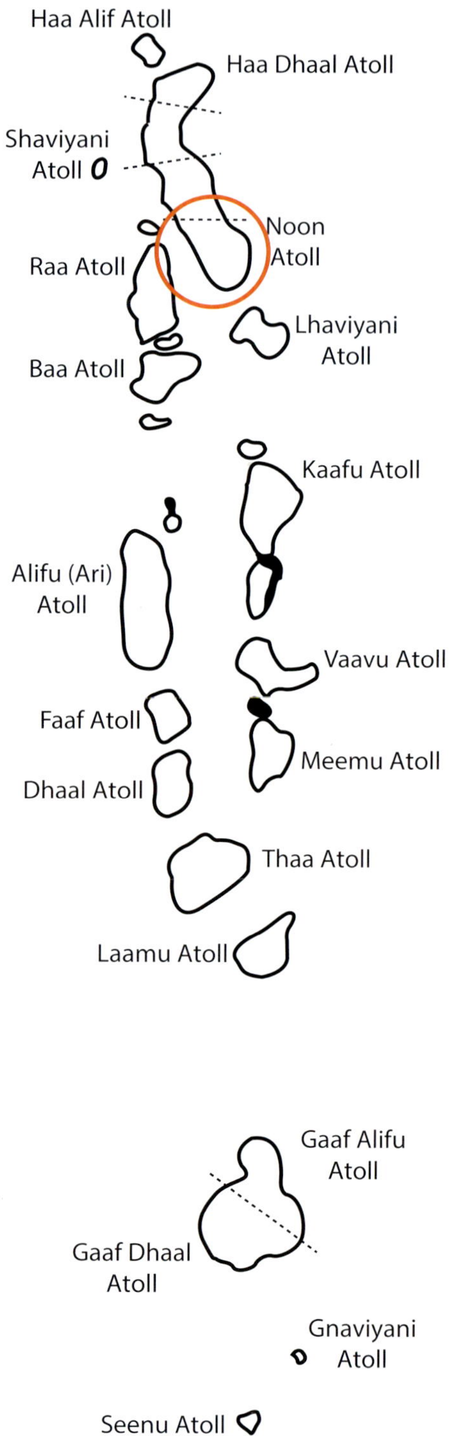

opposite

Location of Meedhoo

right

Location of Meedhoo Old Friday Mosque

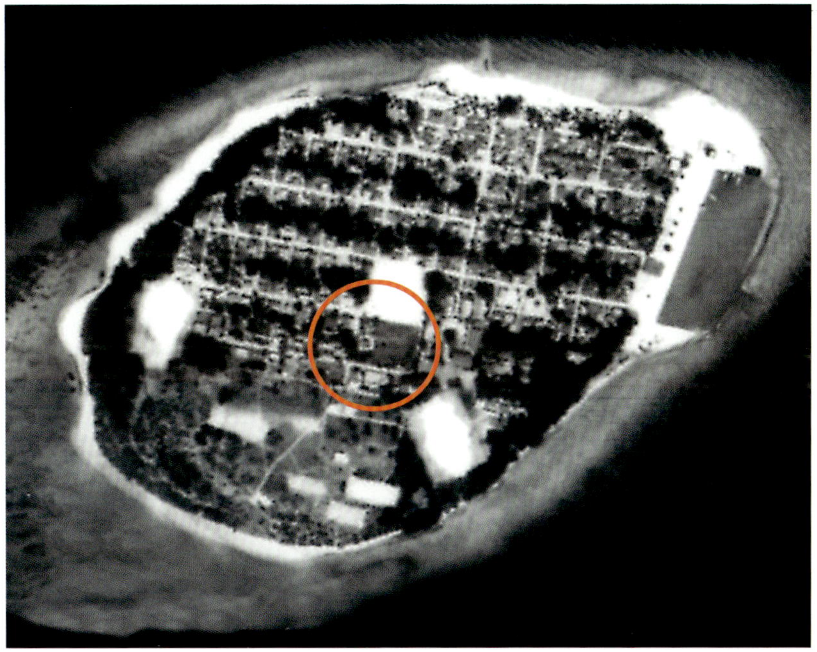

Location

Address: Hiyfaseyha Magu, Meedhoo, Raa Atoll, Maldives

Geographic coordinates: 5° 27' 27.80" N, 72° 57' 16.41"E

Meedhoo is a small island about 154.3 kilometres north of the capital island Male'. It has an area of around 35 hectares and the population is approximately 1,700. The economy of the island depends mainly on tuna and reef fisheries with good fishing in both monsoon seasons. From ancient times, Raa Atoll has been famous for its skilled boat builders, carpenters and fishermen. The atoll was also exposed to early travellers. The legend of Koimala as well as both of the Maldivian journeys of Ibn Battuta began in this atoll. The history of the atoll is linked to a Sayyid from Yemen who settled in Fasmandhoo to become the atoll chief and whose descendants rose to power.

History

c1705 The period of original construction is not clear but according to local oral history it is around 300 years old, probably built during the reign of Sultan Muzaffar Mohamed Imaduddin. The design and construction details of the mosque suggest an even older mosque, possibly 400 years old. Until the 1900s, the mosque roof and interior walls were made from coconut thatch.

1900s The mosque was restored and remodelled extensively in the 1900s. The roofing was changed from coconut thatching to Indian clay roofing tiles, the *dhaala* fitted with timber lattice windows (*miskithu jaali*) and parts of the interiors were replaced. Lacquer workers from Thulhadhoo in the Baa Atoll did the overall lacquerwork and interior columns. Even though it was remodelled, the coral structure and exterior form of the mosque remain unchanged.

1950s The mosque compound had a *haruge'* adjoining the mosque.

1980s The *haruge'* was removed and a boundary wall was built to separate the cemetery.

Description

The mosque compound with mosque, cemetery and wells

Mosque capacity: 80 persons.
Approximate mosque area: 70 square metres.

The mosque compound contains the mosque, two water wells, and a cemetery with a mausoleum. The square coral stone well remains intact, in good condition and is still in use. A large part of the original mosque compound designated to the cemetery is separated with a wall now and has a number of old coral stone tombstones with fine carvings. A coral masonry boundary wall surrounds the mosque with two entrances. Trees such as frangipanis, coconut palms and other flowering plants are present within the complex.

opposite

Existing exterior of the mosque

top

The mosque compound with mausoleum

bottom

Tombstones

opposite and right

Existing stairs

top

Detail of calligraphic carving

opposite

Detail of the stairs and entrance

right

The prayer hall

opposite

Laage'

right

The prayer hall

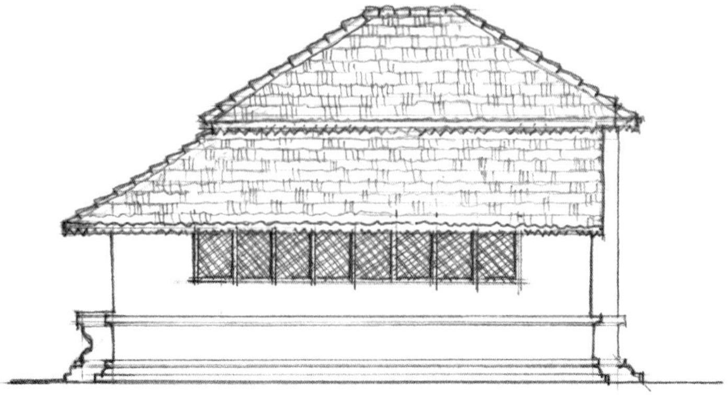

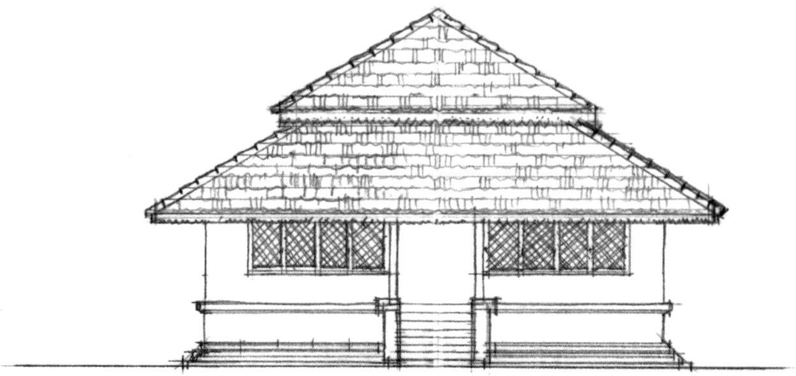

The mosque is one of the few in the country to undergo scarcely any changes since the early 1900s. It is a typical mosque with the prayer hall, mihrab chamber, *dhaala* (antechambers) on three sides and stairs rising to the entrance opposite the mihrab chamber. The mimbar is inside the mihrab chamber. The mosque is built on a coral stone plinth with coral stone walls, timber columns, beams, doors and windows. The roof form is two-tiered and its structure is supported by the integrated walls, columns, beams and doors. The mosque has a timber coffered ceiling with a central recessed *laage'*, raised from the main beams using *thona*.

The exterior coral stone ornament does not have the carvings and finesse of other similar mosques, even if the mouldings and form do. The interior ornament is composed of many timber carvings, lacquerwork and calligraphy. The mosque was renovated in the early 1900s when many parts of the interior were replaced, including the four main columns and a lot of the lacquerwork, which was undertaken by the renowned lacquerworkers of Thulhadhoo. This is the only coral stone mosque still standing with original Indian clay roofing tiles.

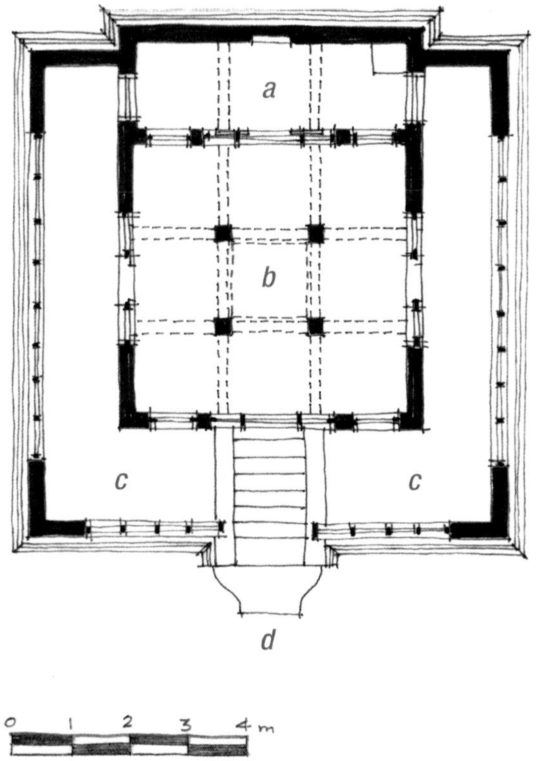

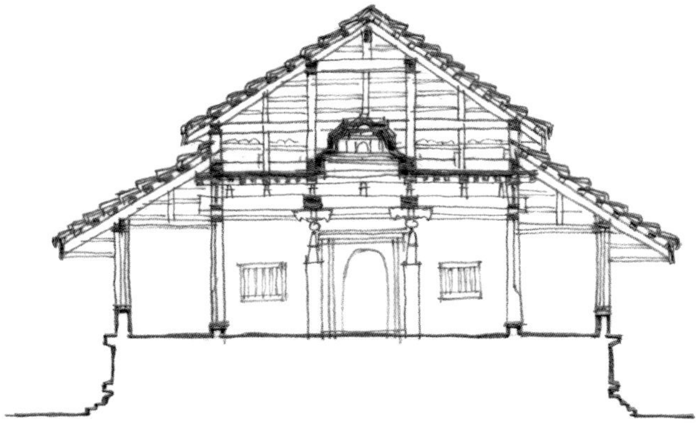

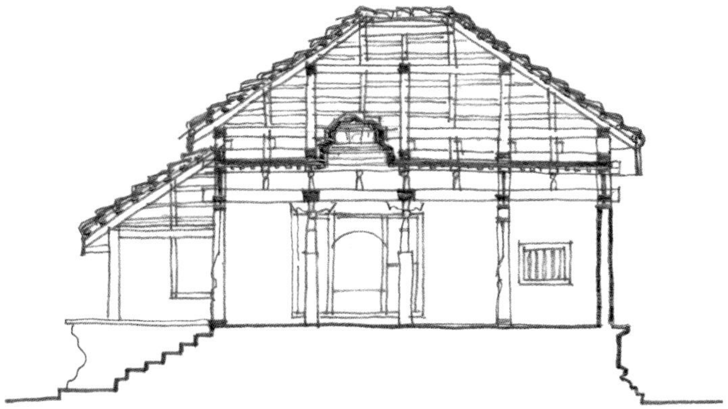

The authors' views of the original mosque

opposite top

Elevations

opposite bottom

Floor plan:

a Mihrab chamber

b Prayer hall

c *Dhaala*

d Entrance

top

Sections

This mosque was at its best after it was restored in the 1900s. It has to be imagined with an Indian clay tiled roof, coral stone wells, tombstones and lacquered interiors. The following illustrations depict the mosque in its former glory.

right

Coral calligraphy carving on the interior wall of the *dhaala*

far right

The compound

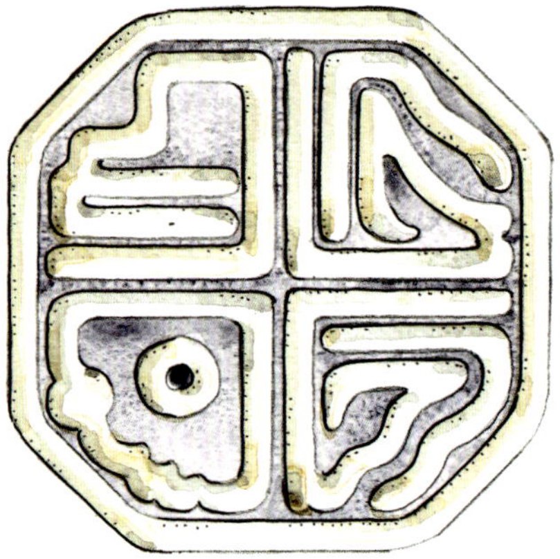

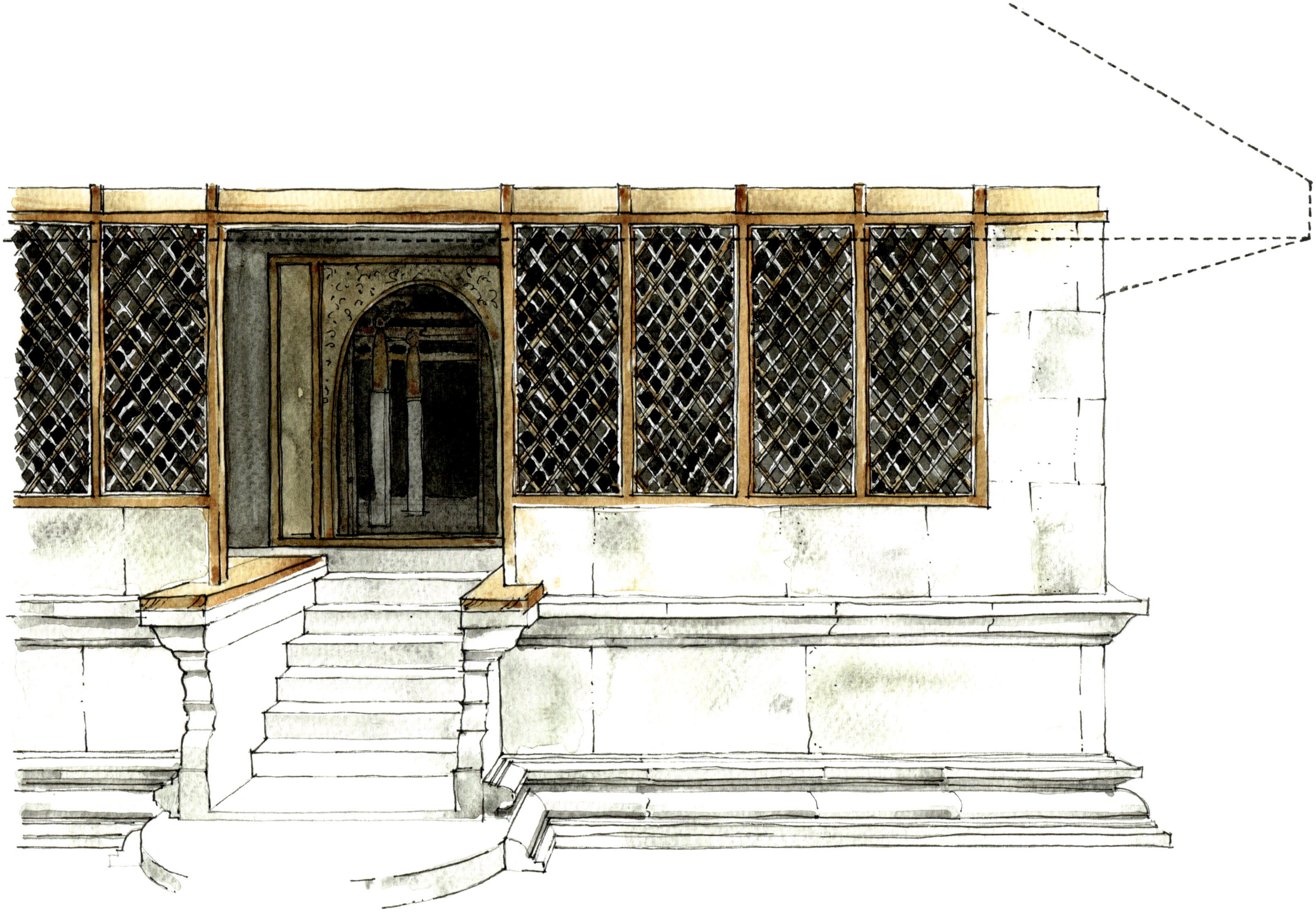

opposite

Exterior entrance detail

right

Interior entrance detail

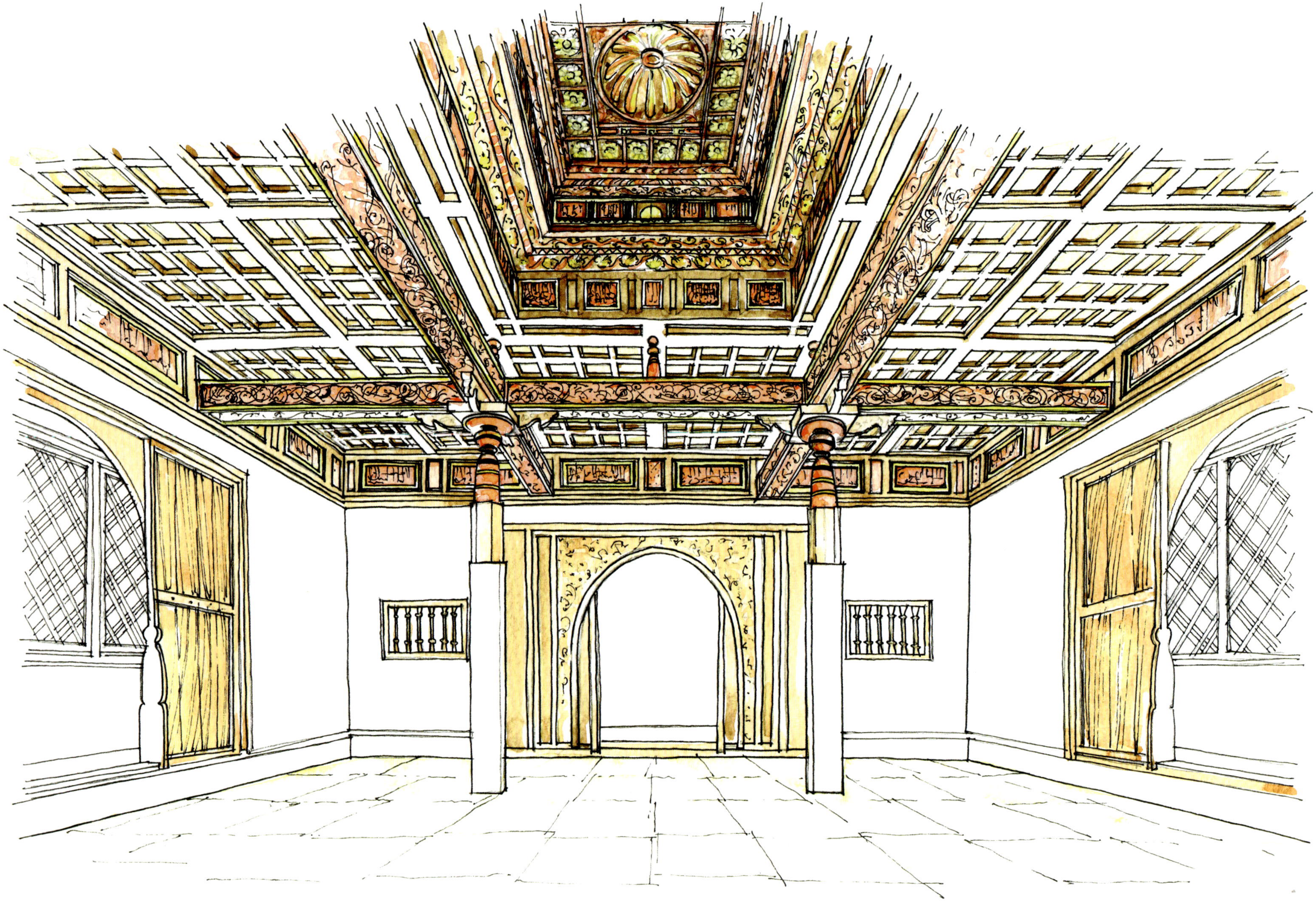

opposite
Interior of the prayer hall
right
Lacquered detail of a timber column head and main beams
bottom
Detail of a lacquered beam

Male’ Kalhuvakaru Mosque, 1789

Male’ Kalhuvakaru or Ebony Mosque is a beautiful small mosque with high-quality coral stone carvings, timber carvings and a lacquered interior. It is the only mosque in which precious ebony timber has been used for its construction.

The mosque and the coral stone well have been relocated and are now in a corner of Sultan Park on the capital island of Male’. It was originally built in 1789 during the reign of Sultan Hassan Nooruddin (1779–1799) and except for one brief period it has been used as a mosque ever since.

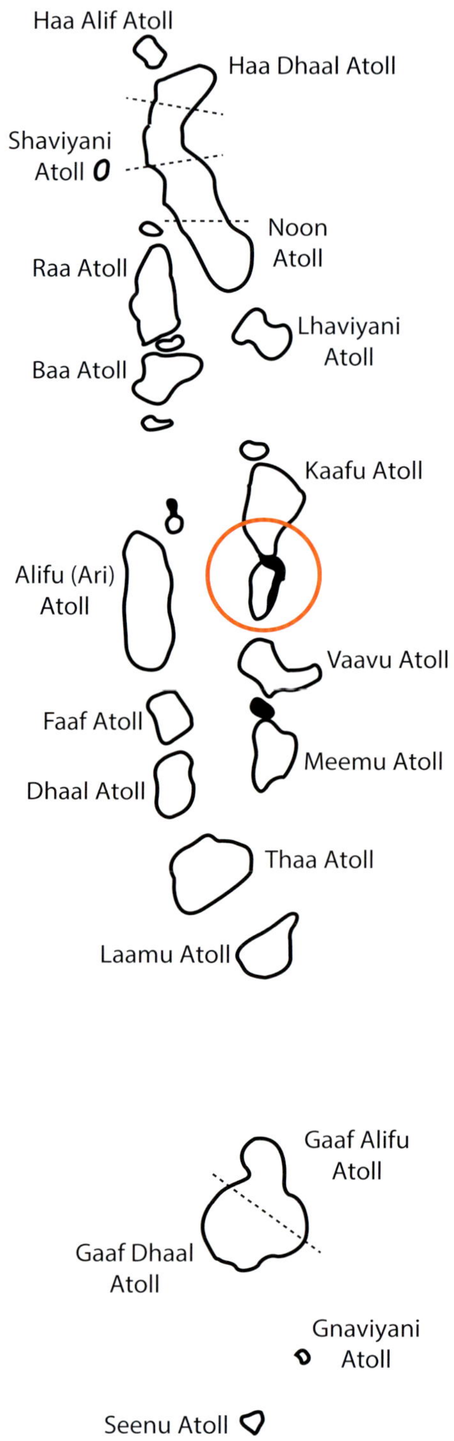

opposite

Location of Male'

right

Location of Male'

Kalhuvakaru Mosque

Location

Address: Shaheed Kuda Adamge
Kuda Thuthu Hingun, Male', Kaafu Atoll,
Maldives
Geographic coordinates: 4° 10' 37.12" N,
73° 30' 40. 49" E

Male' is the capital of Maldives and is about 200 hectares. With a population of more than 100,000 it is among the densest and smallest capital cities in the world. Male' has been the centre for politics, commerce and administration going as far back as the Buddhist period. The original location of Male' Kalhuvakaru Mosque was on the eastern part of the main road, Majeedhee Magu.

History

1789 The mosque was constructed during the reign of Sultan Hassan Nooruddin. It was originally in Majeedhee Magu, in the Henveiru ward. The original roofing was probably coconut thatch.

1970s The mosque was disassembled and moved to Kaafu Furanafushi, a tourist resort in Male' atoll, where it remained unassembled until it was moved back to Male' again.

1980s The mosque was carefully reassembled in its present location, which is a corner of Sultan Park. A metal roof was added to the mosque and remains in good condition.

Description

Mosque capacity: 20 persons.
Approximate mosque area: 24 square metres.

The mosque is now in a corner of the national park of Male'. Since the mosque and its well have been relocated, the present mosque compound has no original features.

The mosque has a prayer hall with *dhaala* (antechambers), on three sides. Opposite the mihrab, on the eastern end, there is an entrance with rising steps. There is no mihrab chamber. Typical for this kind of mosque, Male' Kalhuvakaru Mosque is built on a raised coral stone plinth with coral stone walls, timber doors, beams and roof structure all joined using an interlocking dry joinery system. There are no columns in this mosque and the roof is two-tiered with a modern metal roof covering. There is a highly decorated coffered ceiling with a single stepped recess as a *laage'* raised from the beams by *thona*. The timber used for construction is ebony, one of the hardest

Mosque plinth and entrance

and most precious timbers in the world. The exterior of the plinth is decorated with mouldings and exquisite surface carvings of geometric and organic patterns.
The interior is decorated with organic patterns carved on the doors, a panel of fine lacquerwork and turned timber with calligraphic lacquerwork. With its delicate exterior coral carvings and interiors with carved wooden doors, lacquerwork and calligraphy, the quality of workmanship is some of the best in the country. Originally the floor would have been finished with traditional reed mats on a lime mortar screed and the mosque would have been lit by oil lamps.

opposite

Entrance and stairs

above and far right

Detail of coral carvings on the plinth

right

Detail on the entrance stairs newel post

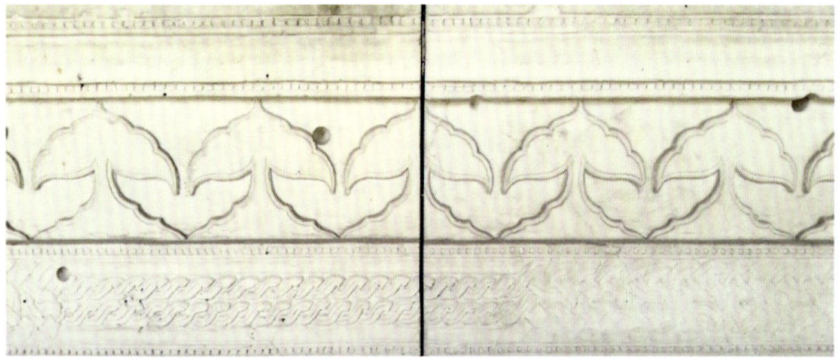

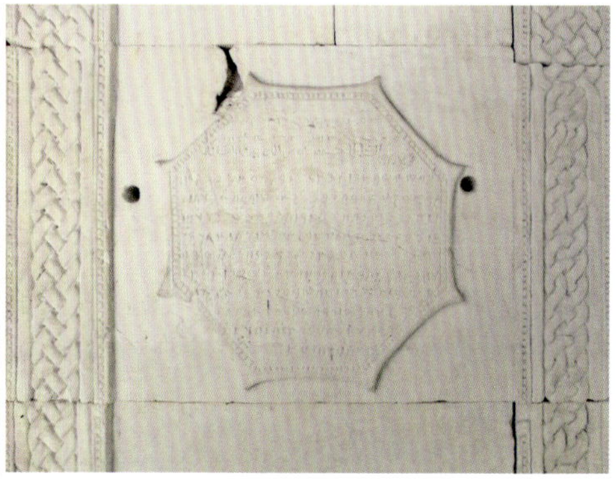

top

Detail of coral carvings on the plinth

right

Stair details

far right

Mosque information carved in coral stone

opposite

The *dhaala*

opposite

Ceiling and *laage'*

right

The prayer hall

far right

Detail of a lacquered beam

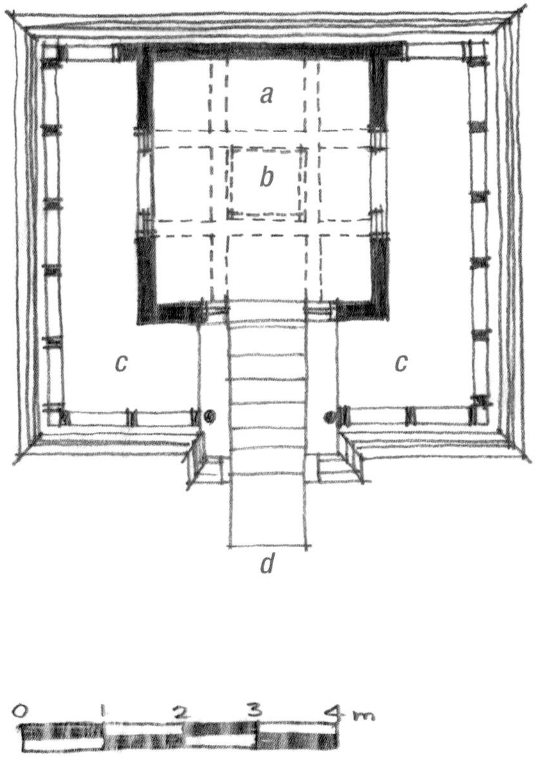
a
b
c
c
d
0 1 2 3 4m

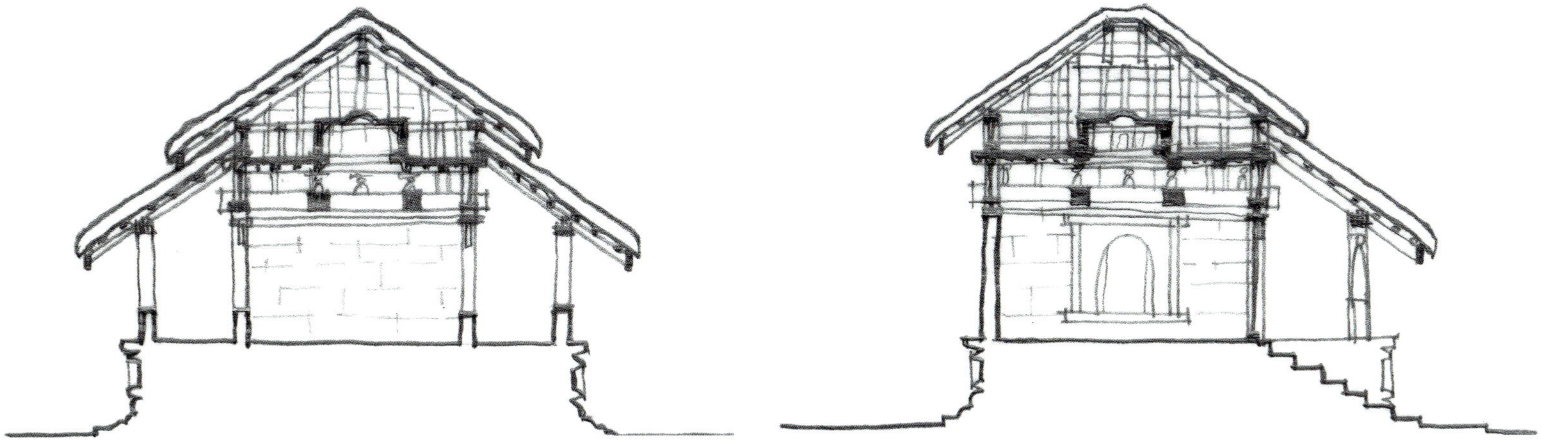

opposite top

Elevations

opposite bottom

Floor plan:

a Mihrab wall

b Prayer hall

c *Dhaala*

d Entrance

top

Sections

far right

Entrance detail

overleaf left

The compound

overleaf top right

Coral carvings – finite organic knot patterns – on the entrance stairs newel post

overleaf bottom right

Coral carvings – bands of organic and geometric interlaced and arabesque patterns – on the plinth

The authors' views of the original mosque

Even though the mosque has been relocated, it can be imagined in the urban setting of Male' with coconut-thatched roofing, coral stone well and tombstones. The interior would have been the same but with traditional reed floor mats. The following illustrations depict how the mosque would have looked in its original state.

opposite

Interior view of the prayer hall

right

Lacquered calligraphy panel on the wall panel of the prayer hall, Jeli Thuluth style

far right

Detail of ceiling, lacquered timber main beam and *Thona*

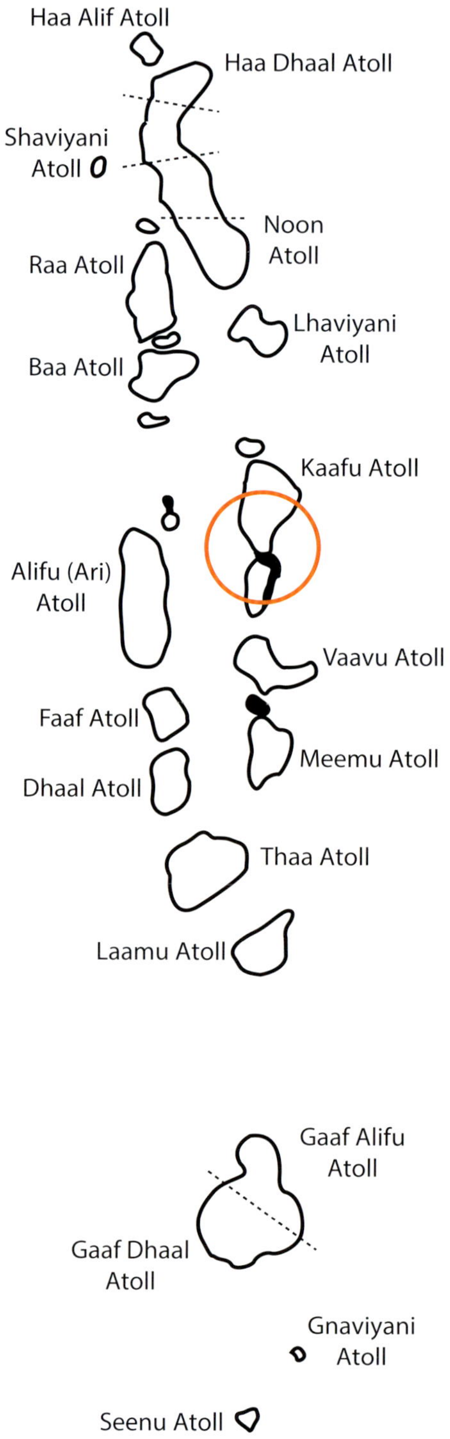

Male' Eid Mosque, 1815

Male' Eid Mosque is one of the most important mosques and the last coral stone mosque to be built, completing the story of coral stone mosques in Maldives. It is extremely beautiful in terms of ornament and workmanship. Its coral stone carvings and decoration are the most elegant among all such work in the world. Unfortunately, over time extensions have been added and the mosque compound has been reduced to simply the mosque and the coral stone well. The old mosque remains in good condition but the original splendour is no longer apparent because of the extensions and urban congestion around it.

At the western end of the capital island of Male', the mosque was built in 1815 by Sultan Mohamed Muinuddin, replacing an older mosque erected during the reign of Sultan Muzaffar Mohamed Imaduddin. Originally built for the Eid prayers, it is no longer the main mosque for Eid prayers but it still serves as one of the most important mosques in Male'.

opposite

Location of Male'

right

Location of Male' Eid Mosque

Location

Address: Handhuvaree Hingun, Male', Kaafu Atoll, Maldives
Geographic coordinates: 4° 10' 29.39" N, 73° 30' 15.29" E

Male' Eid Mosque is at the western end of Male' next to the old Maafannu garrison. During the time of the Sultans, it was the main mosque for Eid prayers and a temporary structure called Eid Haruge' was put up near it to accommodate a huge gathering. The Sultan would lead a grand Eid procession from the palace to the mosque. At present this area of Male' is congested with narrow roads and many high-rise buildings and the ambience of the old setting of the mosque can no longer be appreciated.

History

1815 The mosque was constructed during the reign of Sultan Mohamed Muinuddin (1799–1835) replacing an older mosque built during the reign of Sultan Muzaffar Mohamed Imaduddin (1620–1648).

1940s The original roofing material was probably coconut thatch but there is photographic evidence from the 1940s that Indian clay roof tiles were there for many years of its glorious past.

1970s The mosque was extended to accommodate the growing assembly and the mosque compound was reduced due to land constraints in the area.

2004 Scientific restoration work was carried out to protect the old building from deterioration.

Description

Capacity of the old mosque: 60 persons. Approximate mosque area: 51 square metres.

Owing to urban congestion in the area, the mosque compound has been reduced to the mosque and a coral stone well. The old garrison wall, its gate and the bathing tank no longer exist, so the beauty and glory of the mosque can no longer be appreciated from outside. At present, the

The mosque in its compound

well-preserved old mosque is attached to a modern extension on three sides. It is a typical coral stone mosque with a prayer hall with *dhaala* (antechambers) on three sides and a grand entrance with rising stairs. There is no mihrab chamber and the mimbar is near the corner of the prayer hall. The side *dhaalas* are enclosed and were designated for the Sultan and his entourage.

opposite

The mosque entrance

clockwise

Detail of the mosque plinth; detail of coral carvings on the plinth; sultan's seal; newel post details; mosque stair details

opposite and right

Doorway to the prayer hall

far right

Ceiling and calligraphy panel

opposite and right

The prayer hall

far right

Laage'

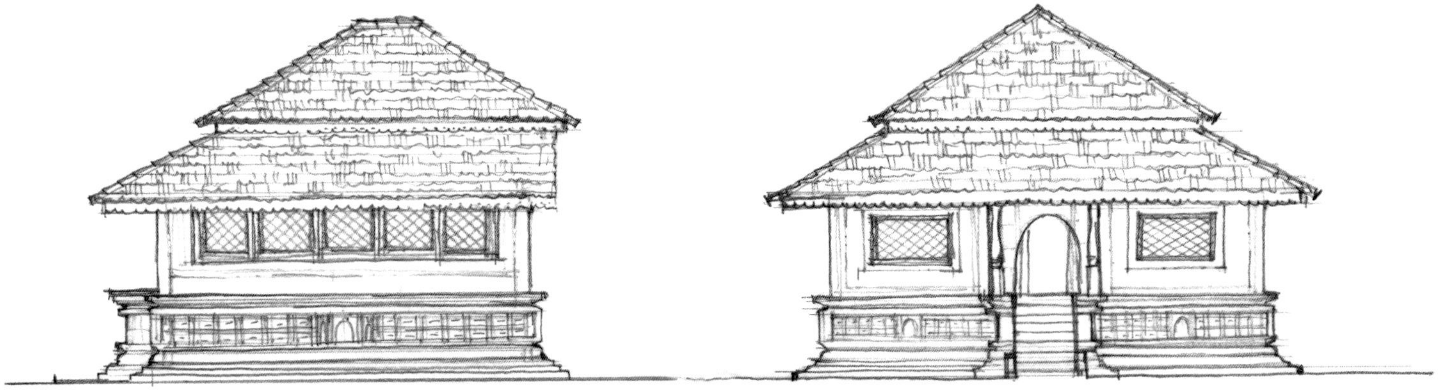

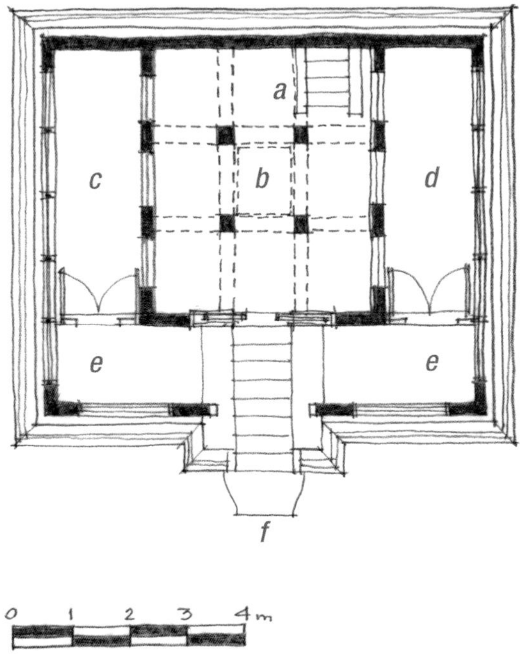
a
b
c
d
e
e
f
0 1 2 3 4m

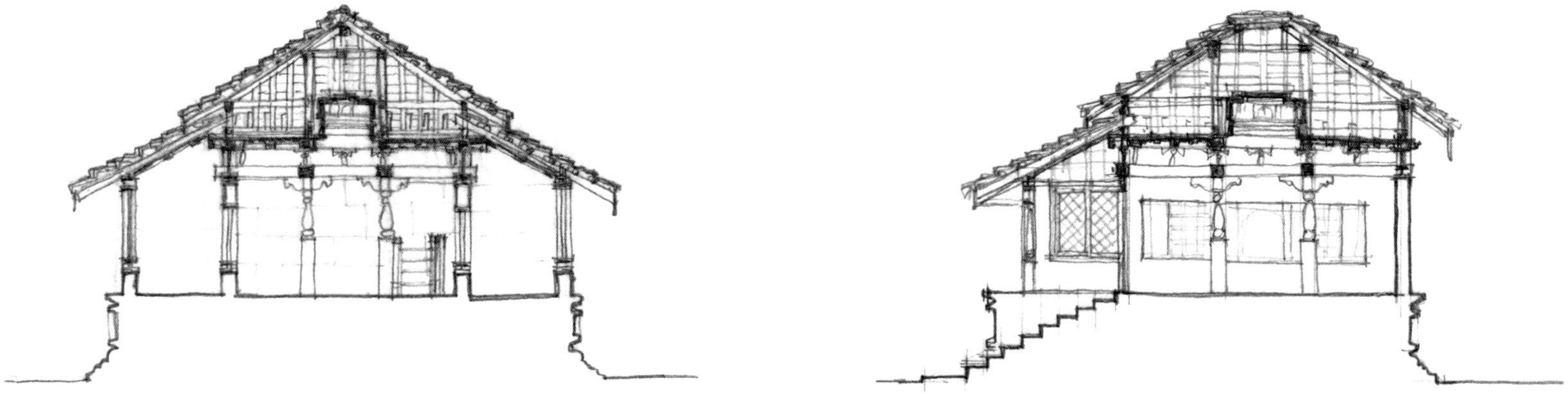

The authors' views of the original mosque

opposite top

Elevations

opposite bottom

Floor plan:

a Mimbar

b Prayer hall

c Sultan's chamber

d Sultan's aides

e *Dhaala*

f Entrance

top

Sections

overleaf

The compound

The mosque is built on a highly decorated coral stone plinth. Coral stone walls, columns, beams, doors and the roof structure are integrated into one structural system by using an interlocking dry joinery system. The roof is two-tiered with a modern metal roofing finish and has a highly decorated coffered ceiling with a central recessed *laage'* raised from the main beams by turned and lacquered *thona*. There are four columns in the prayer hall made from decorated coral stone. This mosque has a unique decorated window system with sliding panels between the side *dhaalas* and prayer hall. With its fine exterior coral carvings and interiors with carved wooden doors, lacquer work and calligraphy, the quality of workmanship is the finest in the country, if not the whole world.

To appreciate the old mosque in its former glory it has to be imagined with Indian clay tile roofing, the coral stone well, the bathing tank and the old garrison wall. The interior would have been the same but lighter and brighter with traditional reed floor mats. The following illustrations depict the mosque in its original state.

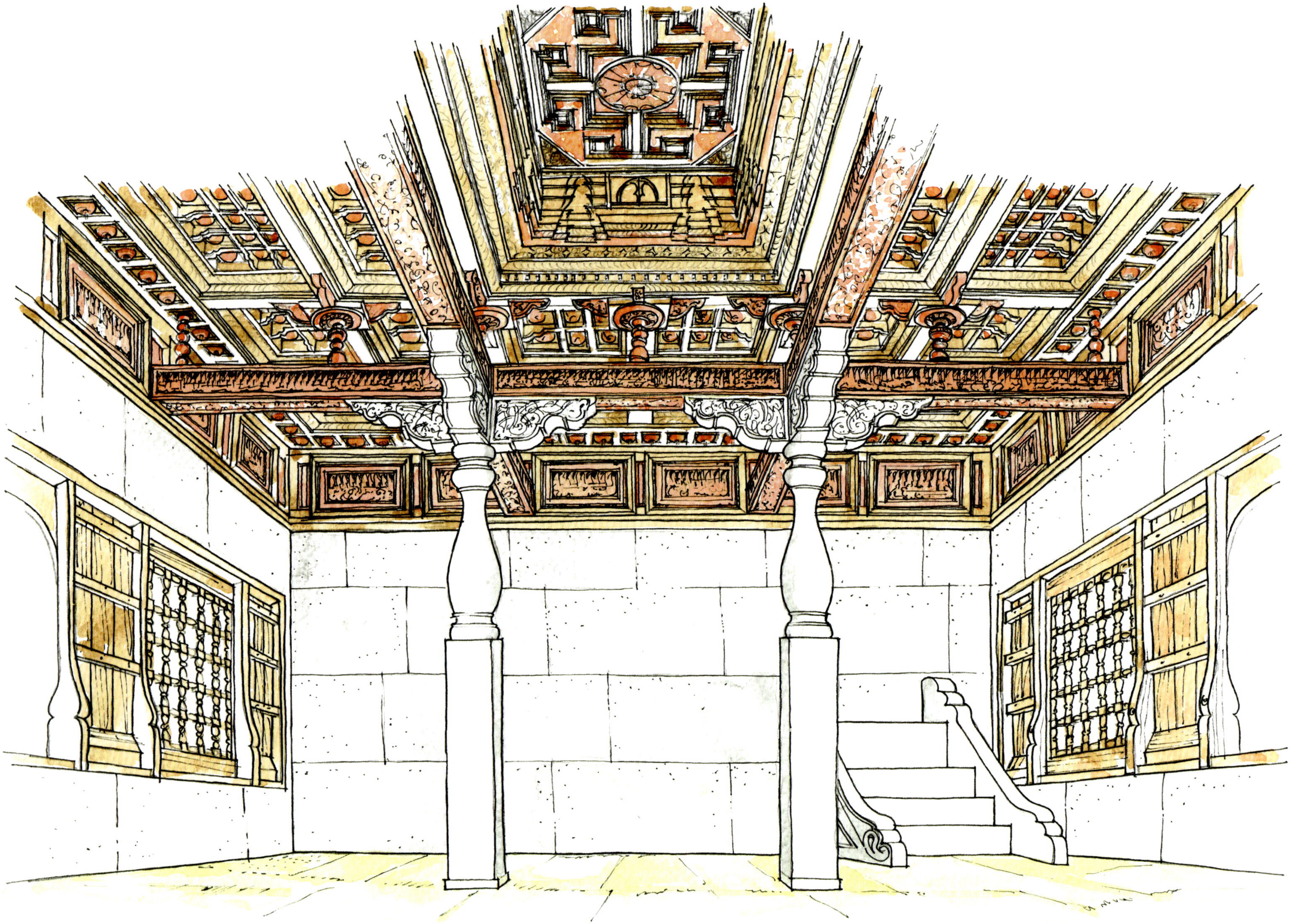

opposite

Interior of the prayer hall

right

Coral carvings – finite organic knot patterns – on the entrance stairs newel post

far right

Lacquered timber entrance stair post

opposite left

Detail of a panel – carved finite palmette pattern – on the timber entrance door to the prayer hall

opposite right

Entrance detail

top

Timber entrance door to the prayer hall

bottom

Lacquered calligraphy panel on the wall panel of the prayer hall, Jeli Thuluth style

Appendix

Historic Periods of Maldives

Ancient period (before 2000 BCE–*c*500 BCE)
The earliest evidence that organised communities and trade existed in Maldives is from 2000 BCE. Originally small communities from the nearby Indian subcontinent must have settled in different parts of Maldives. According to legend, communities from the south of the subcontinent followed by people from the Indus Valley, called Dheyvis, settled in Maldives during this period. A mysterious sea people called Redhin also visited and made temporary settlements.

Classical Hindu and Buddhist period (*c*500 BCE–1153)
Indo-Aryans and other groups from India and Sri Lanka continued to migrate to Maldives introducing the Hindu and Buddhist religions and establishing monastic settlements. Buddhism was introduced around 500 BCE and thrived, with monastic rulers trading with the flourishing traders of the Indian Ocean region. And even though the evidence is still not very clear, Austronesians from Indonesia when migrating to Madagascar also transited and settled in Maldives. Later during this period, Chola kings raided and controlled parts of Maldives.

Medieval Islamic period (1153–1514)
Maldives embraced Islam in 1153 and started to be ruled by dynasties of sultans and sultanas. Trade flourished in the Indian Ocean, as it did in Maldives, uniting people from the whole region. Ibn Battuta, the great Moroccan traveller, wrote the earliest account of Maldives in 1343. Ma Huan, the chronicler of the great Chinese admiral Zheng He, also described Maldives.

Early modern or colonial period (1514–1835)
The Portuguese dominated trade in the Indian Ocean region but after 15 years of control by the Portuguese and their Malabari allies, Ghazi Mohamed Thakurufaanu liberated Maldives. Later in this period Maldives became a free port, trade prospered and many great mosques were built. Francois Pyrard of Laval, a French navigator, wrote detailed accounts of Maldives.

British colonial period (1835–1965)
Maldives became a British protectorate to seek protection from Malabari raiders. The British controlled trade in the whole region and Maldivian trade and travel shifted to British controlled ports. From 1856, Vora traders from Gujarat controlled trade in Maldives. In 1941, the British Royal Navy established a base on Gan Island.

Modern independent period (1965–present)
Maldives became fully independent as a republic in 1965 and Vora traders were expelled. International air travel, shipping, telecommunications, tourism and fisheries changed and continue to change dramatically the political, social and economic landscape of the country. In 2008, a new constitution was ratified taking the form of a modern multiparty democracy.

Brief Timeline of Events in Maldives and the World, 1153–1835

Medieval Islamic Period 1153–1528

Male' or Theemuge' dynasty

1153–1165

Reign of Sultan Mohamed ul- Adil, son of Koimala, known as Dharumavantha Rasgefaanu.
1153: Maldives embraced Islam in the hands of Maulana Abul-Barakaathul Barbaree.
1153: First mosque built on the location of the present Male' Old Friday Mosque. Male' Dharumavantha Rasgefaanu Mosque and the original Nilandhoo, Kudahuvadhoo and Maaenboodhoo mosques were probably built during this period.

1153: The golden period of the Muslim world when Baghdad was the centre for commerce, science, philosophy, medicine and knowledge. Basra flourished as a great port city.
1100–1300: Isfahan Friday Mosque, Iran.

1165–1306

Reign of Sultan Mute' Kalaminja and 14 other sultans, mostly in the name of the Kalaminjas.
1192: Earliest record of the history of Maldives written in copperplate grants.

1169: Saladin defended Cairo from crusaders and founded the Ayyubid Sultanate.
1288: Rise of Islamic sultanates in South East Asia and the decline of the Srivijaya empire of Java.
1271–1295: Travels of Marco Polo.
1184–1199: Koutoubiya Mosque, Morocco.

1306–1347

Reign of Sultan Umar Veeru, father of Rehendhi Khadeeja, and Sultan Ahmad Shihabuddin.
1343: Ibn Battuta wrote the first detailed account of Maldives and its trade in fish, coconuts, coir, cowries, ambergris and tortoise shells with Arabia, India and China.
The original Seedhee Mosque in Male' is believed to have been built by Sultan Umar Veeru.

1300s: Calicut rose to fame as a great port city and the Malabar region became wealthy and powerful.
1325: The beginning of travels by Ibn Battuta.
1300s: East African port Sultanate Kilwa prospered.
The mosque in Kilwa was rebuilt in coral stone.

1347–1388

Reign of Rehendhi Khadeeja, the first sultana who ruled three times, and seven other sultans and sultanas

Hilali dynasty

1388–1420

Reign of Sultan Hilali Hassan I, who seized the throne from Sultan Usman I who came to power by force, and six other sultans.
*c*1397: Fuvah Mulah Kedeyre and Gen mosques.
*c*1400: Guraidhoo Ziyaaraiy Mosque.
*c*1403: Hulhumeedhoo Boadha Mosque.
*c*1417: Hulhumeedhoo Athara Mosque.

1371–1591 The Kotte Kingdom rules the south-west of Sri Lanka and controls the spice trade in Sri Lanka.
1300s Aceh and Samudera Pasai in Sumatra became wealthy port sultanates.
1404: Malacca emerged as an important trading port, becoming an interchange between the Chinese and Indian Ocean trade.
1405–1433: Voyages of Admiral Zheng He.
1333–1391: Alhambra Palace in Granada, Spain.
1356: Mosque-Madrassa of Sultan Hassan, Cairo, Egypt.

1420–1443

Reign of Sultan Dhanna Mohamed, Sultan Yusuf II and Sultan Abu Bakr.
1425: Ma Huan, the Muslim chronicler of Admiral Zheng He visited and wrote about the Maldives.
The original Bandara Mosque, Dhanna Mohamed Rasgefaanu Mosque and Kanbaafaanu Palace in Male' were built by Sultan Dhanna Mohamed.

1443–1491

Reign of Sultan Haji Hassan III and eight other sultans. Haji Hassan brought 70 slaves with him on his return from Hajj. Sultan Sayyid Mohamed, an Arab, ruled while he was on Hajj.

c1400s: Trading activities increased and the Indian Ocean became a highway for travel by different types and sizes of vessels, from the coir rope sewn dhows to the great junks from China.
1429: Mamluk kings increased the prominence of Jeddah as the greatest port city in the Red Sea.
1436: Malacca's ruler converts to Islam.

1491–1528

Reign of Sultan Kalhu Mohamed and five other sultans. Sultan Kalhu Mohamed, a tyrant, ruled three times spanning 32 years until 1528.
1513–1514: Sultan Kalhu Mohamed signed a treaty with the Malabari allies of the Portuguese. The story of Buraki Rani is based on Kalhu Mohamed.

1498: Vasco da Gama arrived in south-west India.
1510–1514: The Portuguese capture Goa, Malacca and reach the Moluccas and Canton to monopolise trade in the region.
1500s: Majapahit empire of Java declines and the Demak sultanate gains power.
c1500s: Demak Great Mosque, Indonesia.

Early Modern or Colonial Period 1528–1835

Hilali dynasty

1528–1554

Reign of Sultan Hassan Shirazi VIII, son of Sultan Kalhu Mohamed and a Persian concubine, followed by his nephews Sultan Mohamed and Sultan Hassan IX.
1529: Visit of the two French travellers, Jean and Raoul Parmentier, who wrote about the navigation skills of Maldivians.
1552: Sultan Hassan IX converted to Christianity and moved to Goa. His efforts to return to power, with the help of Portuguese raiders, and convert Maldivians were defeated by Abu Bakr and Ali (later sultans).

1554–1558

Reign of Sultan Abu Bakr II and Sultan Ali VI.
1556: After several attacks Portuguese raiders martyred Sultan Abu Bakr.
1558: The Portuguese, led by Andhiri Andhirin, invaded Maldives and martyred Sultan Ali VI.

1500s: Swahili cultures of East Africa thrive through trade and slave trade expanded in West Africa.
1542–1552: The Portuguese backed Christian missionary work by Francis Xavier.
c1556: Portuguese dominance and control in the Indian Ocean increases with their trading posts in Africa, the Middle East, India, South East Asia, all the way to Macao, China.
1556: Akbar succeeds the throne of the Mughal and expands the empire to almost all of India north of the Godavari River.
1550–1557: Mosque of Suleiman the Magnificent, Turkey.

1558–1573

Portuguese raiders led by Andhiri Andhirin controlled Maldives.
*c*1565: Three Utheemu brothers and Ali Haji Hassan Haji led a freedom struggle, organising their incursions from the southern atolls.

Utheemu dynasty

1573–1585

Reign of Sultan Ghazi Mohamed Thakurufaanu.
1573: After a struggle of eight years led by Mohamed Thakurufaanu and supported by Ali Raja, Maldives was liberated from the Portuguese.
Vaadhoo Dhanna Kaleyfaanu returned from Yemen.
*c*1500s: Original Utheemu Kandhuvalu Mosque, Kolhufushi Ghazi, Baarah and Dhidhdhoo Eid mosques and the original Vaadhoo Old Mosque.

c1558: The Ottomans led by Suleiman the Magnificent challenged Portuguese and Spanish dominance in the Indian Ocean.
c1564: The Portuguese link Indian Ocean trade via the Philippines to the Americas.
1570: Japan opens trade in Nagasaki.
1571: The Spanish capture Manila and trade through Mexico.
1581: Union of Spain and Portugal.
1571–1575: Friday Mosque, Fatehpur Sikri, India.

1585–1620

Reign of Sultan Ibrahim III (Kalaafaanu), son of Sultan Ghazi Mohamed, and Sultan Husain Faamudheyri Kilegefaanu II.
1609: After disputes, Ali Raja of Cannanore suddenly raided Male'. Kalaafaanu was martyred in battle at sea and his heir Kalhuthukkala was taken away.
1602–1607: Francois Pyrard of Laval, a French navigator, wrote a detailed account of Maldives.
*c*1586: Hulhumeedhoo Fandiyaru Mosque, original Bihurozukamana Mosque, Hangnaameedhoo and Dhevvadhoo old mosques.

1588: The British defeat the Spanish Armada and become rulers of the Atlantic.
1600: Start of the British East India Company.
1602: Start of the Dutch East India Company.
1612: First British trading factory set up in Surat, India.
1619: The Dutch establish a trading post in Batavia (Jakarta).

1620–1648

Reign of Sultan Shujai Mohamed Imaduddin I, known as Kalhuthukkala.
1624–1625: A formidable Portuguese expedition to regain power failed. In their retreat they burned down the mosque in Villingili. Later the sultan fortified Male' and bought guns from Aceh. Failed revolt by a noble, Samiya Kilege, who set up a separatist state in Maafilaafushi.
*c*1620: First Male' Eid Mosque (rebuilt later), Veyvah Old Mosque and first Olhuvelifushi Old Mosque.

1622: Iran expels the Portuguese from Hormuz.
1627–1658: Rule of Shah Jahan in India.
1640: The Dutch break Portuguese control in the Indian Ocean and the British set up a trading factory in Madras.
1641: The Dutch seize Malacca from the Portuguese.

1648–1687

Reign of Ibrahim Iskandhar I, son of Kalhuthukkala.
1649: Expedition by the Portuguese with the grandsons of Christian King Hassan IX to gain power was defeated. From this time paying annual tribute to Ali Rajas was stopped and after many defeats the Rajas gave up sending expeditions.
Male' became a free port and trade improved with ships from many parts of the region.
1658: Present Male' Old Friday Mosque;1668: Mosque gates; 1675: Minaret, and Kondey Mosque.
Mathiveri Old Mosque was probably built during this period.

1658: The Dutch controlled Sri Lanka and traded in Maldives.
1663: The Dutch take control of Cochin and the Malabar region and are more tolerant to different religions than the Portuguese.
1668: Bombay was leased to the British.
1674: Start of the French East India Company in Pondicherry, India.
1648: Taj Mahal was completed.

1687–1692

Reign of Sultan Kuda Mohamed, Ibrahim I and Sultan Mohamed Muhiddin ul-Adil.
Sultan Kuda Mohamed, who was six years old, and his mother Mariyam the concubine, who poisoned his father Sultan Iskandhar Ibrahim I, gained absolute power and led an indecent life. They both met a fateful death.

1691: The British set up a trading factory in Calcutta, India.

Early Modern or Colonial Period 1528–1835 continued

Isdhoo dynasty

1692–1704

Reign of Sultan Shamsuddin I, a descendant of the Quraish who reigned for five months and dedicated himself to Islam; Sultan Mohamed of Dhevvadhu, a former Qazi and the first sultan of the Isdhoo dynasty, and three other sultans. Sultan Ibrahim Muzhiruddin was the last sultan of the dynasty. This was a prosperous period and many mosques were built and repaired.
1692–1701: Construction of Friday mosques in Nolhivaram, Kelaa, Ihavandhoo, Isdhoo; two mosques in Dhevvadhoo; reconstruction of Fenfushi and Kudahuvadhoo Friday mosques and renovation of Male' Dharumavantha Rasgefaanu Mosque.
Bandidhoo Old Mosque and original Fuvah Mulah Old Friday Mosque probably built during this period.

c1700s Decline of the Persian Safavid empire.
c1700s The British improve trade ties with Canton, China.

Dhiyamigil dynasty

1704–1721

Reign of Sultan Muzaffar Mohamed Imaduddin II. Sultan Imaduddin II, former Chief Wazir, ascended the throne after infighting due to mistaken news that the former sultan died during his Hajj trip. The deposed sultan later made several unsuccessful attempts to regain power.
c1705 Construction of Meedhoo Old Friday Mosque.

1707 Decline of the Mughal empire ravaged by civil wars. The French and British take advantage.
1717 The Mughal empire allows the British East India Company to trade without duties.

1721–1752

Reign of Sultan Ibrahim Iskandhar II and Sultan Mohamed Imaduddin III. A peaceful period trading with the Dutch in Sri Lanka.
1726: Hassan Thaajuddin compiled a history of Maldives.
1729–1742: Male' had many fires, cyclones and famine.
1752: Ali Raja raided and kidnapped Sultan Mohamed Imaduddin III (who died in captivity) and the son of Sultan Iskandhar II, and looted and torched the palace and Male'.

Cannanore Ali Rajas became powerful by forming an alliance with the Sultans of Mysore.
1735: The French took control of Mahé in Malabar and Pondicherry in east India.

1752–1759

Ali Raja of Cannanore controlled Maldives for five and a half months. Maldives was liberated by Dhon Bandaara (later Sultan Ghazi Hassan Izzuddin), with the help of the French in Pondicherry, and was governed by Askarun (soldiers) until Bandaara's enthronement.

1757: Robert Clive's victory in the Carnatic Wars against Indian-French allies marks the beginning of British rule in India.

Huraa dynasty

1759–1767

Reign of Sultan Ghazi Hassan Izzuddin. This sultan was known to be wise and just to his people and relatives of Sultan Iskandhar II.
1761: Ali Raja's expedition of eight vessels was defeated.
1763: Son of Iskandhar II escaped Cannanore fort and returned to Male' through Ganjam.
First Huraa Old Mosque.

1760: Beginning of the Industrial Revolution.
1761: Defeat of the French in the Indian Ocean by the British.

Dhiyamigil dynasty

1767–1773

Reign of Sultan Mohamed Ghiyasuddin.
The sultan, son of Sultan Iskandhar II a skilled warrior, compiled his chronicles of war and the first Thaana dictionary.
1771: Ali Raja's expedition of nine vessels was defeated.
When the sultan departed for Hajj a revolution brought the brother of Ghazi Hassan Izzuddin to power.

1770: Landing of Captain Cook in Australia.
1773: Warren Hastings becomes the first British Governor and resides in Calcutta.

Huraa dynasty

1773–1779

Reign of Sultan Mohamed Shamsuddin II and Sultan Mohamed Muizzuddin.
1774: Sultan Ghiyasuddin when returning from Hajj was seized and secretly drowned. Ali Raja's expedition led by Kakaa Thakurufaanu was defeated.
1774: Construction of Male' Kalhuvakaru Mosque commenced.

1775–1785: The British defeated in the Anglo-Maratha War in India.

1779–1799

Reign of Sultan Haji Hassan Nooruddin.
1797: Two shipwrecks provided the sultan with treasures and weapons.
1799: The sultan dies from illness during his second Hajj, suffered after grievous conflicts with Sharif Ghalib of Mecca who raided his ship and camp. He disregarded advice by elders not to travel with treasures.
1789: Completion of Male' Kalhuvakaru Mosque.

1786: British settlements in Penang.
1788: British settlements in Australia.
1796: The British defeat the Dutch in Sri Lanka.

1799–1835

Reign of Sultan Mohamed Muinuddin
1818–1819: A great famine occurred.
1828: Maldives appealed to the British in Colombo to assist in defusing hostilities with the Ali Rajas.
1835: Beginning of British involvement with permission for the hydrographic mapping of Maldives by Captain Moresby of the British Royal Navy.
1815: Male' Eid Mosque rebuilt (the last coral stone mosque).

1802: Sri Lanka became a British colony.
1807: Abolition of the slave trade begins.
1815: The British seize Cape Colony, South Africa.
1819: British settlements in Singapore.

Glossary

ambaa
repetitive chants or shanties

atoll
group of coral islands

bandhi
rhyming poetry or chants

beru dhunmaari thaalhafili
drums and trumpets for royal festivals and processions

boduberu
traditional double-headed drums

bokkura
a light rowing boat of around 4.5 metres used near the shore

coir rope
rope made from coconut fibre

Cypraea moneta
scientific name for the cowry shell used as money in ancient times

dhaala
veranda-like antechamber attached to the prayer hall and stairs of mosques

dhaani
a container attached to a pole used for drawing water from wells

Dhathuru dhoni and *odi*
traditional sailboats of around 15 metres used for open sea inter-island travel and cargo

Dhathuru battheli and *bandu odi*
large sailboats of around 22 metres used for open sea long-distance travel and cargo

dhoni
traditional Maldivian boat

dhoni haruge'
boat house

dikka
raised tribune in mosque for Quran or prayer recitals

Dhives
early Maldivian script

Eveyla
earliest known Maldivian script

farihi
ballads

fenda
veranda-like antechamber

fuh hiri and *oman hiri*
smooth-grained types of Porite coral stone

goh hiri
a knotty-grained type of Porite coral stone

haruge'
building structures for purposes other than living

havaru beru
drums for public gathering

hirigaa
Porite coral or boulder coral

hiri galu vadaan
coral stone carpentry

holhu ashi
small beach pavilion

Jeli Thuluth
a calligraphic variety of Arabic script

kasabu
traditional embroidery

Kulhi jehun
display of traditional martial arts

laafenkurun
traditional lacquerwork on a flat surface

laage'
stepped recessed part in the centre of the mosque ceiling

langiri and *bandiyaa jehun*
traditional music with rhythmic dance movements

liyelaajehun
traditional lacquerwork on turned timber

madhaha and *mauloodh*
religious songs dedicated to the Prophet Mohamed (PBUH)

mas dhoni and *mas odi*
traditional sailboats of about 9 metres used for open sea fishing

mihrab
a niche in the wall of the mosque that faces towards Mecca

mimbar
pulpit in the mosque

miskithu jaali
timber window lattice used in mosques

miskiy
mosque

onu hiri
a coarse-grained type of Porite coral stone

Porites
type of coral stone

qiblah
point toward which Muslims turn to pray

raivaru
sung poetry with swapped letters

rihaakuru
a thick fish paste made in Maldives

Thaana
current Maldivian script

thaara
traditional single-headed drum

thelhigaa
chipped coral stone

thona
short posts supporting the ceiling from the main beams

veligaa
coral sandstone

Bibliography

Agrawal, OP. 'Conservation of Mosques and Museum Objects', UNESCO Technical Report FMR/CLT/86/103 (Paris), 1986.

Ahmad, Y and Jameel, MM 'Coral Stone Mosques of Maldives – Towards World Heritage Nomination: Phase 1 of Preparation for a World Heritage Serial Nomination: Final Report', Dept of Heritage (Male'), 2012.

Archnet. Archnet: Digital Library: Coral Stone, 2010. Retrieved 20.07.2012, from http://archnet.org/search/summary.jsp?search_id=1696570

ASI. Archaeological Survey of India; Monuments, 2012. Retrieved 23.03.2012, http://asi.nic.in/asi_monuments.asp.

Bell, HCP. *The Maldive Islands, An Account of the Physical Features, Climate, History, Inhabitants, Production and Trade*, Ceylon Government Press (Colombo), 1883.

Bell, HCP. *The Maldive Islands, Monograph on the History, Archaeology and Epigraphy*, Ceylon Government Press (Colombo), 1940.

Bose, S. *A Hundred Horizons: The Indian Ocean in the Age of Global Empire*, Harvard University Press (Cambridge, MA and London), 2009.

Carswell, J. 'Mosques and Tombs in The Maldive Islands', Art and Archaeology Research Paper 9, June 1976.

Chaudhuri, KN. *Trade and Civilisation in the Indian Ocean: An Economic History from the Rise of Islam to 1750*, Cambridge University Press (Cambridge, UK), 1985.

Didi, MA. *Minicoy: Seventy Miles Away from Maldives* [written in Dhivehi] (Male'), MA Didi (Male'), 1979.

Didi, MI. *Maldivians in the Opinion of Christopher* [written in Dhivehi], National Centre for Linguistic and Historic Research (NCLHR) (Male'), 1981.

Didi, MI. *From Pyrard's Pen* [written in Dhivehi], NCHLR (Male'), 1995.

Fee, CV. *The Encyclopedia of Malaysia: Architecture*, Archipelago Press (Kuala Lumpur), 1998.

Forbes, ADW. 'The Mosque in the Maldive Islands: A Preliminary Historical Survey', *Archipel*, 26(1), 43–74, 1983.

Frishman, M and Khan, H (eds). *The Mosque: History, Architectural Development and Regional Diversity*, Thames and Hudson (London), 2007.

Heyerdahl, T. *The Maldive Mystery*, Adler & Adler (Bethesda, MD), first edition, 1986.

Horton, M. 'Primitive Islam and Architecture in East Africa', *Muqarnas*, vol 8, 1991.

Isaacs, RB and Blurton, TR. *Visions from the Golden Land: Burma and the Art of Lacquer*, Art Media Resources (Chicago), 2000.

Jameel, MM. 'Architectural Typological Study of Coral Stone Mosques of Maldives', MSc Thesis, University of Malaya, Malaysia (2012).

Kirkman, J. 'Mnarani of Kilifi: The Mosques and Tombs', *Ars Orientalis*, vol 3, 95–112, 1959.

Lambourn, E. 'Carvings and Communities: Marble Carvings for Muslim Patrons at Khambhat and around the Indian Ocean Rim, Late Thirteenth–Mid-Fifteenth Centuries', *Ars Orientalis*, vol 34, 99–133, 2004.

Luthfee, M. *Introduction to Geography of Maldives* [written in Dhivehi], Luthfee (Male'), 1995.

Mackintosh-Smith, T. *The Travels of Ibn Battutah*, Picador (London), 2002.

Maldives-NCLHR. *Malé Hukuru Miskiiy*. NCHLR (Malé), 1986.

Maloney, C. *People of the Maldive Islands*, Orient Longman (Bombay), 1980.

Maniku HA & Wijayawardhana, GD. *Isdhoo Loamafaanu*, Royal Asiatic Society of Sri Lanka (Colombo), 1986.

McKillop, H, Magnoni, A, Watson, R, Ascher, S, Winemiller, T, & Tucker, B. 'The Coral Foundations of Coastal Maya Architecture: Research Report in Belizean Archaeology', Proceedings from 2003 Belize Archaeology Symposium, Belize, 2004.

Michell, G. *Architecture of the Islamic World: Its History and Social Meaning*, Thames & Hudson (London), 2009.

Mikkelsen, E. *Archeological Excavation of a Monastery in Kaashidhoo*, NCLHR (Male'), 2000.

Mohamed, N. *Dhivehi Writing Systems*. NCLHR (Male'), 1999.

Mohamed, N. *Heritage Places of Maldives* [written in Dhivehi], NCLHR (Male'), 2007.

Mohamed, N. *Essays on Early Maldives*, NCLHR (Male'), 2008.

Mohamed, N. *Maldivian Travellers* [written in Dhivehi], Haveeru (Male'), 2012.

Naseer, A. 'Profile and Status of Coral Reefs in Maldives and Approaches to its Management', Proceedings from FAO Regional Workshop on the Conservation and Sustainable Management of Coral Reefs, 1997.

National Centre for Linguistic and Historic Research (NCLHR), *History of Maldives* [written in Dhivehi], NCLHR (Male'), 1981.

Pearson, M. *The Indian Ocean (Seas in History)* (1 ed), Routledge (London), 2003.

Peterson, A. *Dictionary of Islamic Architecture*, Routledge (London), 1996.

Pyrard, F, Gray, A. *The Voyage Of Francois Pyrard V1: Of Laval To The East Indies, The Maldives, The Moluccas and Brazil* (trans), Kessinger Publishing (Whitefish, MT), 1619.

Rasdi, MTM. *Rethinking Islamic Architecture*, Strategic Information and Research Development (SIRD) (Malaysia), 2010.

Ragupathy, P and Mohamed N. *An Etymological Dictionary of Maldivian Island Names*, NCLHR (Male'), 2008.

Reynolds, CHB. 'The Mosques in the Maldives', *Archipel*, vol 28, 1984.

Shafeegu, A. *Maldivian Craftsmanship* [written in Dhivehi], NCLHR (Male'), 1988.

Shafeegu, A. *Boat Building* [written in Dhivehi], NCLHR (Male'), 1991.

Sheriff, A. *Dhow Cultures of the Indian Ocean: Cosmopolitanism, Commerce and Islam*, C Hurst & Co (London), 2010.

Shokoohy, M. *Muslim Architecture of South India: The Sultanate of Ma'bar and the Traditions of the Maritime Settlers on the Malabar and Coromandel Coasts (Tamil Nadu, Kerala and Goa)*, Routledge (London), 2003.

Silva, R. 'Conservation of the Mogul Mosques and the Excavations of Maldivian Mounds', UNESCO technical report FMR/CLT/CH/85/119 (Paris), 1985.

UNESCO-WHC. Preparing World Heritage Nominations, 2011a. Retrieved 20.09.2013, from http://whc.unesco.org/en/activities/643/

UNESCO, WHC. UNESCO World Heritage Tentative List: Historic City of Ahmadabad, Gujarat, India, 2011b. Retrieved 2.10.2011, from http://whc.unesco.org/en/tentativelists/5616/.

UNESCO, WHC. UNESCO World Heritage Site: Group of Monuments at Hampi, Karnataka, India, 2011c. Retrieved 10.11.2011 from http://whc.unesco.org/en/list/241.

UNESCO, WHC. UNESCO World Heritage Site: Ancient City of Polonnaruwa, Sri Lanka, 2011d. Retrieved 15.10.2011, from http://whc.unesco.org/en/list/201.

UNESCO, WHC. UNESCO World Heritage Site: Lamu Old Town, Kenya, 2011e. Retrieved 10.11.2011, from http://whc.unesco.org/en/list/1055.

UNESCO, WHC. UNESCO World Heritage Site: Old City of Sana'a, 2011g. Retrieved 11.04.2011, from http://whc.unesco.org/en/list/385.

UNESCO, WHC. UNESCO World Heritage Site: Ruins of Kilwa Kisiwani and Ruins of Songo Mnara, 2011h. Retrieved 10.10.2011, from http://whc.unesco.org/en/list/144/.

UNESCO, WHC. UNESCO World Heritage Site: Stone Town of Zanzibar, 2011i. Retrieved 10.11.2011, from http://whc.unesco.org/en/list/173.

UNESCO-WHC. World Heritage List, 2013a. Retrieved 28.09.2013, from http://whc.unesco.org/en/list/

UNESCO-WHC. Tentative Lists, 2013b. Retrieved 15.10.2011, from http://whc.unesco.org/en/tentativelist

Vilgon, L. *Maldive and Laccadive Islands Bibliography*, L Vilgon (Stockholm), 1990.

Index

Figures in italics indicate illustrations.